CAMBRIDGE LIBRARY COLLECTION

Books of enduring scholarly value

Life Sciences

Until the nineteenth century, the various subjects now known as the life sciences were regarded either as arcane studies which had little impact on ordinary daily life, or as a genteel hobby for the leisured classes. The increasing academic rigour and systematisation brought to the study of botany, zoology and other disciplines, and their adoption in university curricula, are reflected in the books reissued in this series.

The Natural History of Birds

Georges-Louis Leclerc, Comte de Buffon (1707–88) was a French mathematician who was considered one of the leading naturalists of the Enlightenment. An acquaintance of Voltaire and other intellectuals, he work as Keeper at the Jardin du Roi from 1739, and this inspired him to research and publish a vast encyclopaedia and survey of natural history, the ground-breaking *Histoire Naturelle*, which he published in forty-four volumes between 1749 and 1804. These volumes, first published between 1770 and 1783 and translated into English in 1793, contain Buffon's survey and descriptions of birds from the *Histoire Naturelle*. Based on recorded observations of birds both in France and in other countries, these volumes provide detailed descriptions of various bird species, their habitats and behaviours and were the first publications to present a comprehensive account of eighteenth-century ornithology. Volume 8 covers domestic and foreign marine birds.

Cambridge University Press has long been a pioneer in the reissuing of out-of-print titles from its own backlist, producing digital reprints of books that are still sought after by scholars and students but could not be reprinted economically using traditional technology. The Cambridge Library Collection extends this activity to a wider range of books which are still of importance to researchers and professionals, either for the source material they contain, or as landmarks in the history of their academic discipline.

Drawing from the world-renowned collections in the Cambridge University Library, and guided by the advice of experts in each subject area, Cambridge University Press is using state-of-the-art scanning machines in its own Printing House to capture the content of each book selected for inclusion. The files are processed to give a consistently clear, crisp image, and the books finished to the high quality standard for which the Press is recognised around the world. The latest print-on-demand technology ensures that the books will remain available indefinitely, and that orders for single or multiple copies can quickly be supplied.

The Cambridge Library Collection will bring back to life books of enduring scholarly value (including out-of-copyright works originally issued by other publishers) across a wide range of disciplines in the humanities and social sciences and in science and technology.

The Natural History of Birds

From the French of the Count de Buffon

VOLUME 8

COMTE DE BUFFON
WILLIAM SMELLIE

CAMBRIDGE UNIVERSITY PRESS

Cambridge, New York, Melbourne, Madrid, Cape Town, Singapore,
São Paolo, Delhi, Dubai, Tokyo, Mexico City

Published in the United States of America by Cambridge University Press, New York

www.cambridge.org
Information on this title: www.cambridge.org/9781108023054

© in this compilation Cambridge University Press 2010

This edition first published 1793
This digitally printed version 2010

ISBN 978-1-108-02305-4 Paperback

T H E

NATURAL HISTORY

O F

B I R D S.

FROM THE FRENCH OF THE

COUNT DE BUFFON.

———

ILLUSTRATED WITH ENGRAVINGS;

AND A

PREFACE, NOTES, AND ADDITIONS
BY THE TRANSLATOR.

———

IN NINE VOLUMES.

V O L. VIII.

L O N D O N:

PRINTED FOR A. STRAHAN, AND T. CADELL IN THE STRAND;
AND J. MURRAY, Nº 32, FLEET-STREET.

M DCC XCIII.

CONTENTS

OF THE

EIGHTH VOLUME.

———

A 2

The

CONTENTS.

FOREIGN

CONTENTS.

§

C O N T E N T S.

CONTENTS.

1. The

CONTENTS.

THE

THE

NATURAL HISTORY

OF

B I R D S.

The I B I S *.

OF all the fuperftitious practices that have
ever degraded the human race, the worfhip of animals might be deemed the moft abject
and the moft abfurd : and yet did that propenfity originate from the pureft of motives. In
the early ages of the world, man was on all fides
encompaffed by dangers, and had to ftruggle
naked and unarmed againft the formidable attacks of his numerous foes. Thofe animals,
therefore, which confpired with his efforts to
deftroy and eradicate the hoftile tribes, were na-

---

* In Greek Ιβις, which the Romans adopted. It has no name
in European languages, as being unknown in our climates. According to Albertus, it was called in Egyptian *Leheras*. In Avi-
cenna, the word *Anfchuz* denotes the Ibis. St. Jerome was miftaken in tranflating *Janfchuph (Leviticus,* ii. *Ifaiah,* xxxiv.) by
Ibis, for a nocturnal bird is meant in that paffage. Some inter-
preters render the Hebrew word *Tinfchemet* by Ibis.

turally

turally entitled to his regard and affection. But the fentiment of gratitude afterwards degenerated into veneration ; and fear and intereft, nourifhing the groveling propenfion, both the ufeful and the pernicious creatures were alike exalted into the rank of gods.

Egypt is one of thofe countries where animal-worfhip was of the higheft antiquity, and obferved with the moft fcrupulous attention, for many ages; and that humiliating fpecies of idolatry, which is authenticated by all the monuments that have been tranfmitted to pofterity, feems to prove, that the original fettlers had long contended with the noxious animals. In fact, crocodiles, ferpents, grafshoppers, and all the other loathfome creatures, teemed in the deep and fpacious mud, deluged by the annual inundation of the river. The heat of a tropical fun foftering the rich flime would engender infinite numbers of offenfive and fhapelefs beings, which would fucceffively be effaced, till the earth, purged of its impurities, was occupied by nobler inhabitants.

" Swarms of little venomous ferpents," the early hiftorians relate *, " rofe out of the flime " of marfhes, and flying in a great body towards " Egypt, would have entered into that country " and fpread defolation, had not the Ibis op- " pofed itfelf to their inroad, and repelled them."

* Herodotus, *Euterpe*, N° 76. Ælian, Solinus, Marcellinus, Pomponius Mela, *lib*. iii. 8.

Was

Was not this the fource of the fuperftitious veneration paid to that bird? The priefts encou-
raged the notions of the vulgar; when the gods,
they faid, deigned to affume a vifible form, it
was that of the Ibis. Their tutelar deity *Thoth*
or Mercury, the inventor of arts and of laws,
had already undergone that transformation *;
and Ovid, faithful to this ancient mythology, in
the battie of the gods and giants, conceals Mercury under the wings of an Ibis, &c †. But fet-
ting afide all thefe fables, we have ftill to exa-
mine the hiftory of the combats between thefe
birds and the ferpents. Herodotus affures us,
that he went to view the field of battle. " Near
" the town Butus," he fays, " on the confines of
" Arabia, where the mountains open into the
" vaft plain of Egypt, I there faw immenfe
" heaps of ferpents' bones ‡." Cicero cites this
paffage §, and Pliny feems to confirm it, by faying, that the Egyptians invoke the Ibifes againft
the invafion of ferpents ‖.

We read alfo in the hiftorian Jofephus, that
when Mofes made war on the Æthiopians, he
carried, in cages of *papyrus*, a great number of
Ibifes, to oppofe them to the ferpents ¶. This

---

* Plato in *Phædr.*
† Metam. *lib.* v.
‡ Herodotus, *Euterpe*, Nᵒˢ 75 and 76.
§ Lib. i. *De Nat. Deorum.*
‖ Hift. Nat. Lib. x. 28.
¶ *Antiq. Judaic.* lib. ii. 10.

4   I B I S.

ftory, which is not very probable, is eafily ex-
plained by a fact mentioned by Maillet, in his
defcription of Egypt : " A bird named *Pharaoh's*
" *capon* (known to be the Ibis) follows more than
" an hundred leagues the caravans in their
" route to Mecca, for the fake of the dung left
" at the encampments, though at other times it
" is never feen on that track *." We may
prefume, that the Ibifes thus accompanied the
Hebrew nation in their march out of Egypt;
and that Jofephus has disfigured the fact, by af-
cribing to the prudence of the general what
was due only to the inftinct of the birds; and
has introduced the army of Æthiopians and the
cages of papyrus to embellifh his narration,
and to exalt our idea of the legiflator of the
Jews.

To kill the Ibis was, among the Egyptians,
forbidden under pain of death †. That people, whofe temper was equally gloomy and
vain, invented the lugubrious art of preparing
mummies, by which they endeavoured, we may
fay, to perpetuate death, and to counteract the
benevolent views of nature, which, in compaf-
fion to our feelings, labours affiduoufly to efface
every difmal and funereal image. Not only
were they folicitous to preferve human bodies,
they applied their fkill in embalming to the
facred animals. Many receptacles of mum-

* Defcription de l'Egypt, *partie* II. p. 23.
† Herodotus, *uti fupra.*

mies

mies which have been dug up in the plain of Saccara are called *bird-pits*, becaufe only birds are found embalmed, particularly the Ibis, contained in tall earthen pots, whofe orifice is ftopped with cement. We have received feveral of thefe veffels; and in all of them we difcovered a fort of doll, formed by the bandages which encafed the bird, of which the greateft part fell into black duft when the ligatures were removed. We could however perceive all the bones of a bird, with the feathers fticking to fome bits of flefh that remained folid. From thefe fragments we could judge of the fize of the bird, which was nearly equal to that of the curlew; and the bill, which was preferved in two of the mummies, fhowed the genus: it was as thick as that of a ftork, was curved like the bill of the curlew, but not channelled: and as its curvature is equal throughout, we may place the Ibis between the ftork and the curlews *. In fact, fo nearly is it related to both thefe genera of birds, that the modern naturalifts have ranged it with the latter, and the ancients had claffed it with the former. Herodotus has diftinctly characterized the Ibis, by faying that " its bill is " much hooked, and its legs like thofe of the " crane." He takes notice of two fpecies: " The firft," he relates, " is entirely black; " the fecond, which conftantly occurs, is all

* See one of the bills reprefented by Edwards, *plate* 105.

B 3   " white,

" white, except the tips of the feathers of the
" wing and tail, which are very black ; and the
" neck and head, which are only covered with
" fkin."

But I muft here remove the obfcurity with
which this paffage of Herodotus has been in-
volved by the ignorance of tranflators, and
which cafts an air of fable and abfurdity on the
whole. A claufe which ought to have been
rendered literally, " *which oftener occur among*
" *men's feet*," runs thus in their verfions, " *thefe*
" *indeed have feet like men.*" Naturalifts, at a
lofs to conceive the import of this odd comparifon, have ftrained to explain or palliate it.
They fuppofe that Herodotus miftook the ftork
for the white ibis, and imagined its flat toes to
refemble thofe of a man. But this interpreta-
tion was unfatisfactory; and the Ibis with human
feet might have been rejected among the fables.
Yet under this abfurd image was it admitted as
a real exiftence ; and we cannot help being furprized to find at prefent this account inferted in
the memoirs of a learned academy * : though
the chimera is only the production of the tranf-
lator of that ancient hiftorian, whofe candour in
acknowledging the uncertainty of his narratives,
when drawn from other information, ought to

* " The other fpecies (the White Ibis) has its feet fafhioned
" like the human feet." *Memoires de l'Academie des Infcriptions &*
Belles Lettres, tome ix. p. 28.

procure

procure him credit in fubjects that came under his own obfervation.

Ariftotle, too, difcriminates two fpecies of Ibis; he adds, that the white kind is fpread over all Egypt, except near Pelufium, where only the black ones occur, which are feen in no other part of the country *. Pliny repeats this particular obfervation †. But all the ancients, at the fame time that they remark the difference of the two birds in point of colour, afcribe to them both the fame common figure, habits, and inftincts; and regard Egypt, in exclufion to every other country, as their proper abode ‡. If it was carried abroad, they alledge, it languifhed out its days, confumed by the defire of revifiting its native foil §. A bird fo ardently attached to its country, naturally became the emblem of it: the figure of the Ibis, in the hieroglyphics, denotes Egypt, and few images or characters are oftener repeated on all the monuments. They appear on moft of the obelifks; on the bafe of the ftatue of the Nile, at the Belvidere in Rome, and alfo in the garden of the Thuilleries at Paris. In the medal of Adrian, where Egypt appears proftrate, the Ibis is placed at her fide; and this bird is figured with an

* Hift. Animal. *lib.* ix. 27.
† Hift. Nat. *lib.* x. 30.
‡ Strabo places them alfo on a frefh-water lake, near Lichas, in the extremity of Africa.
§ Ælian.

elephant

elephant in the medal of Quintus Marius, to
fignify Egypt and Lybia, the fcenes of his ex-
ploits, &c.

If fuch was the popular and ancient regard
paid to the Ibis, it is not furprizing that its hiftory has been charged with fables. It has been
faid to procreate with its bill * : Solinus feems
not to doubt this; but Ariftotle juftly ridicules
the notion of virgin purity in this facred bird †.
Pierius relates a wonder of an oppofite kind;
he fays that, according to the ancients, the ba-
filifk was hatched from an Ibis' egg, formed in
that bird from the venom of all the ferpents
which it devoured. They have alfo afferted
that the crocodiles and ferpents, when touched
with an Ibis' feather, remained motionlefs as if
enchanted, and often died on the fpot. Zoro-
after, Democritus, and Philo have advanced
thefe tales; and other authors have reprefented
it as living to an extreme age : the priefts of
Hermopolis pretended even that it might be
immortal, and as a proof they fhowed Appion
an Ibis fo old, they faid; that it was no more
fubject to death.

Thefe are but part of the fictions on the fubject of the Ibis, fabricated in the religious land
of Egypt : fuperftition ever runs into extremes;
but if we confider the political motives that

* Ælian.
† De Generatione Animalium, lib. iii. 6.

would

would induce a legiflator to eftablifh the worfhip of ufeful animals, we muft admit the neceffity in that country of preferving and multi-
plying them, in order to reprefs or extirpate the
noxious tribes. Cicero·* remarks judicioufly,
that no animals were held facred by the Egyptians but fuch as merited regard from extreme
utility to them : an † opinion moderate and
wife, very different from the fentence of the
fevere and violent Juvenal, who reckons the
veneration paid to the Ibis among the crimes of
Egypt; and inveighs againft that worfhip, which
fuperftition no doubt overftrained, but which
prudence ought to maintain; fince fuch is the
weaknefs of man, that the moft profound law-
givers have made that fpurious paffion the foundation of their ftructures.

But to confider the natural hiftory of the Ibis,

---

* *Ægyptii nullam belluam, nifi ob aliquam utilitatem quam ex eâ
caperent, confecrârunt velut Ibes, maximam vim ferpentium conficiunt,
cum fint aves excelfæ, cruribus rigidis, corneo procercque roftro; aver-
tunt peftem ab Ægypto, cùm volucres angues, ex vaftitate Lybiæ,
vento Africo invectas, interficiunt atque confumunt, ex quo fit ut illæ
nec morfu vivæ neceant nec odore mortuæ; eam ob rem invocantur ab
Ægyptiis Ibes.*—De Nat. Deor. *lib.* i.

M. Perrault has miftaken the latter part of this fentence. *Anciens Memoires de l'Academie,* tom. iii. partie 3.

† We can fcarce give this as the reafon of the worfhip of the
crocodile : but that animal had adoration paid it only in a fingle
city of the Arfinoïte tribe, while its antagonift, the ichneumon,
was venerated all over Egypt. Befides, in this city of crocodiles,
thefe deftructive animals were worfhipped under the impreffion of
fear, with the idle view to detain them from vifiting a place whither the ftream naturally never bore them.

we

we find it has a strong appetite to feed on ser-
pents, and even a sort of antipathy to all reptiles.   Belon assures us that it continues to kill
them, though sated with prey.   Diodorus Siculus says, that night and day the Ibis, walking
by the verge of the water, watches reptiles,
searching for their eggs, and destroying the
beetles and grashoppers which they meet.
Accustomed to respectful treatment in Egypt,
these birds advanced without fear into the midst
of the cities.   Strabo relates, that they filled the
streets and lanes of Alexandria to such a degree
as to become troublesome and importunate, con-
suming indeed the filth, but also attacking provisions, and defiling every thing with their
dung: inconveniencies which would shock the
delicate and polished Greek, though the Egyptians so grosly superstitious, might cheerfully
submit to them.

Thefe birds breed on the palm-trees, and
place their nest in the thick bunches of the
sharp leaves, to be safe from the attacks of their
enemies, the cats *.   It appears that they lay
four eggs, such at least is the number which
we may infer from the explication given by
Pignorius of the table of Isiacus : He says,
that the Ibis " makes its eggs after the manner
" of the moon † ;" which seems to have no

* Philo, *de propriet. Animal.*
† *Ad lunæ rationem ova fingit.*   Menf. Ifid. Explic. p. 76.

other

other import than what Dr. Shaw has noticed, that the bird lays as many eggs as the moon has phafes. Ælian explains why the Ibis was con- fecrated to the moon, and marks the time of its incubation, by faying that it fat as many days * as the ftar Ifis took to perform the revo- lution of its phafes †.

Pliny and Galen afcribe the invention of the clyfter to the Ibis, as they do the letting blood to the hippopotamus ‡ : " Nor are thefe the " only things," the former adds, " in which " man has profitably imitated the fagacity of " animals §." According to Plutarch, the Ibis ufes only falt-water for that purpofe. Perrault, in his anatomical defcription of this bird, af- ferts that he obferved a hole in the bill, through which the water might be difcharged.

We have faid that the ancients diftinguifhed

---

* Plutarch affures us that the young Ibis, juft hatched, weighs two drachms. *De Ifid. & Ofir.*

† Clement of Alexandria, defcribing the religious repafts of the Egyptians, fays, that among other difhes they carried round among the guefts an Ibis; this bird, by the black and white of its plu- mage, was the emblem of the dark and lucid moon. *Stromat.* lib. v. p. 671. And, according to Plutarch *(De Ifid. & Ofir.)* the lunar crefcent was reprefented by the difpofition of the white upon the black of the plumage.

‡ Galen. *lib. de Plebot.*

§ *Simile quiddam (folertiæ hippopotami, fibi junco venam aperi- entis) & volucris in eadem Egypto monftravit, quæ vocatur Ibis; roftro aduncitate per eam partem perluens, quâ reddi ciborum onera max- ime falubre eft. Nec hæc fola multis animalibus reperta funt ufui futura & homini.* Plin. *lib.* viii. 26.—Alfo Plutarch, *De Solert.*

two

two species of Ibis, the white and the black. We have seen only the white; and though Perrault says that the black ibis is oftener brought to Europe than the white, no naturalist has seen it since Belon, from whom we must give the description.

THE EGYPTIAN IBIS.

The WHITE IBIS.

Tantalus-Ibis. Linn. and Gmel.
Ibis Candida. Briff.
The Emfeefy or Ox-bird. Shaw.
The Egyptian Ibis. Lath.

T H I S bird is fomewhat larger than the curlew, and fomewhat fmaller than the ftork :
its length from the point of the bill to the end
of the nails is about three feet and an half.
Herodotus defcribes it as having tall naked
legs; the face and front equally deftitute of
feathers ; the bill hooked; the quills of the tail
and wings black, and the reft of the plumage
white. To thefe characters we fhall add fome
other properties not mentioned by the ancient
hiftorian : The bill is rounded, and terminates
in a blunt point; the neck is of an equal thick-
nefs throughout, and not clothed with pendant
feathers like that of the ftork.

Perrault defcribed and diffected one of thefe
birds, which had lived in the *menagerie* at Ver-
failles. He found, on comparing it with a
ftork, that it was fmaller, but its bill and feet
proportionally longer ; that the feet of the ftork
were only four parts of the whole length of the
§   bird,

bird, while thofe of the Ibis were five parts.
He obferved the fame proportional difference
to obtain between their bills and their necks.
The wings appeared very large; their quills
were black, and all the reft of the plumage was
white, inclined a little to rufty, and diverfified
only by fome purple and reddifh fpots under the
wings; the top of the head, the orbits, and the
under fide of the throat, were void of feathers,
but covered by a red wrinkled fkin; the bill
was thick at the root, round, an inch and half
in diameter, and curved the whole length; it
was of a light yellow at its origin and deep
orange near the extremity: the fides of the bill
were fharp, and fo hard that they might cut
ferpents *, which is probably the way that the
bird takes to deftroy them; for the tip being
blunt it could fcarce pierce them.

The lower part of the legs was red, and meafured more than four inches; though Belon, in
his figure of the black ibis, reprefents it as only
one inch in length; both that part and the
foot were entirely covered with hexagonal
fcales. The fcales which incrufted the toes
were cut into tablets, and the nails were pointed, ftraight, and blackifh. Both fides of the
mid-toe were bordered by the rudiments of a
membrane, which, in the two other toes, appeared only on the infide.

---

* *Corneo proceroque roftro.* Cicero, *uti fupra.*

Though

Though the Ibis is not granivorous, its ventricle is a fort of gizzard, whofe inner membrane is rough and wrinkled. We have more than once remarked this incongruity in the ftructure of birds; in the caffowary, for inftance, which does not feed on flefh, the ftomach is membranous like that of the eagle *.

Perrault found the inteftines to be four feet eight inches long; the heart was of a middling fize, and not extremely large, as Merula pretended; the tongue, which was very fhort, and concealed at the bottom of the bill, was only a fmall cartilage invefted by a flefhy membrane; which gave occafion to Solinus' remark, that this bird had no bill. The globe of the eye was fmall, not exceeding fix lines in diameter. " This White Ibis," fays Perrault, " and ano-" ther which was kept at the *menagerie* at Ver-" failles, both of them brought from Egypt, were " the only birds of this kind ever feen in France." According to him, all the defcriptions of the modern authors have been borrowed from the

---

* An interefting circumftance in this defcription concerns the paffage of the chyle in the inteftines of birds. Injections were made into the mefenteric vein of one of the ftorks diffected with the Ibis, and the liquor paffed into the cavity of the inteftines : and a portion of inteftine having been filled with milk and tied at both ends, the compreffed liquor paffed into the mefenteric vein. Perhaps, adds the anatomift, this paffage is common to all the tribe of birds; and as they exhibit no lacteal veins, we may juftly conjecture, that this is the courfe of the chyle in paffing from the inteftines into the mefentery.

ancients.

ancients. This remark appears to be juft; for Belon did not recognize the White Ibis in Egypt, which is improbable, if he had not taken it for a ftork.

[A] Specific character of the White Ibis, *Tantalus-Ibis*: " Its face is red, its bill yellow, its feet gray, its wing-quills black, its body tawny-whitifh."

The BLACK IBIS.

Tantalus Niger. Gmel.
Numenius Holosericeus. Klein.
Ibis Nigra. Charleton.

" THIS bird," says Belon, " is somewhat smaller than a curlew * ;" it is smaller therefore than the white ibis, and must also be shorter : yet the ancients assert that the two species were similar in every respect, except in colour. The present is entirely black ; and Belon seems to insinuate that the front and face are covered with bare skin, by saying that the head is like that of a *cormorant*. But Herodotus, who seems to have bestowed attention on his two descriptions, does not represent the head and neck as featherless. The other characters and the habits are stated to be the same in both birds.

* " This Black Ibis is as high on legs as a bittern, and its bill is as thick as the thumb at its origin, pointed at the end, vaulted, and something curved, entirely red, as are the thighs and the legs." *Observ. de Belon*, Paris 1555. liv. ii. p. 102.

[A] Specific character of the Black Ibis, *Tantalus Niger.* " Its face, its bill, and its feet, red ; its body black."

The C U R L E W S.

LES COURLIS. *Buff.*

FIRST SPECIES.

Scolopax-Arquata. Linn. and Gmel.
Numenius. Briff. Will. Klein, &c.
* *Numenius Arquata.* Lath.

THOSE words which imitate the cries of animals are the names affigned them by nature, and are the firſt which men have impoſed. The ſavage languages exhibit innumerable examples of theſe inſtinctive appellations, and they have been more or leſs preſerved in the poliſhed tongues ; in the Greek eſpecially, the fineſt and the moſt deſcriptive. Without the name *elorios*, the ſhort deſcription which Ariſtotle gives of the Curlew would be inſufficient to diſtinguiſh it from other birds †. The French

---

* In Greek Ελωριος, or Νεμηνιος : in Latin *Numenius, Arquata, Falcinellus* : in Italian *Arcaſe, Torquato* : in German *Wind-Vogel, Wetter-Vogel, (wind-bird, weather-bird)* ; and on the Rhine, near Straſburg, *Regen-Vogel (rain-bird)* : in Dutch *Hanikens* : in Daniſh *Heel-Spove*, and *Regen-Spaaer* : in Norwegian *Lang-neel, Spue* : in Lapponic *Guſgaſtak.*

† " The *elorios* is a bird that lives near the ſea, and like the rail ; it feeds along the ſhore in fine weather."

THE COMMON CURLEW.

names *courlis*, *turlis*, are words imitative of its voice *; and in other languages, the appellations *curlew*, *caroli*, and *tarlino*, &c. mark the same relation. The epithets *arquata* and *falcinellus* allude to the hooked form of its bill †: and so also does the term *numenius*, derived from *neomenia* or new moon; because the bill resembles the moon's crescent. The modern Greeks denominate it *macrimiti*, or long nose ‡, on account of the great length of its bill compared with that of the body. The bill is slender, furrowed, equally curved throughout, and terminated in a blunt point; it is weak, and its substance tender, and calculated only to dig up the worms from the soft earth. This character might set the Curlew at the head of a numerous tribe of birds, such as the woodcocks, the snipes, the horsemen, &c. which, not being armed with a bill fit for catching or piercing fish, are obliged to subsist on the various insects and reptiles that swarm in mud and in wet boggy grounds.

The neck and feet of the Curlew are long; the legs partly naked, and the toes connected near their junction by a portion of membrane. The bird is nearly as large as a capon; its total length about two feet; that of its bill five or six inches; its alar extent more than three feet.

---

* Belon.
† Gesner. He gives the same derivation of the Italian *Arcase*.
‡ Belon.

Its

Its whole plumage is a mixture of light gray, except the belly and rump, which are entirely white; dafhes of brown are interfperfed over all the upper parts, and each feather is fringed with light gray or rufty; the great quills of the wing are of a blackifh brown *; the feathers of the back have a filky glofs; thofe of the neck are downy, and thofe of the tail, which fcarce extends beyond the wings, are, as well as the middle ones of the wing, interfected with white and blackifh brown. There is little difference between the male and the female †, which is only fomewhat fmaller ‡; and therefore the particular defcription which Linnæus has given of it § is fuperfluous.

Some naturalifts have afferted, that the flefh of the Curlew has a marfhy tafte; but it is much prized, and ranked by feveral with that of the water fowls ‖. The Curlew lives on earth-worms, infects, periwinkles, &c. which it gathers on the fea-beach, or in the marfhes and wet meadows ‖: its tongue is very fhort, and

---

* On account of the mottled plumage of the Curlews, Schwenck-feld terms them *pardales*; but unfortunately for the refinements of nomenclature, that name would rigoroufly exclude more than half of the fpecies of Curlews.

† Belon.

‡ Willughby.

§ *Numenius Rudbeckii.* Fauna Suecica, N° 139.

‖ Willughby and Belon.

‖ *Idem.* Willughby fays, that he once found a frog in its ftomach.

concealed

concealed at the bottom of the bill. Small
pebbles *, and fometimes grain †, are found in
its ftomach, which is mufcular like that of the
granivorous birds ‡. The *œfophagus* is inflated
like a bag, and overfpread with glandulous *papillæ* §. There are two *cæca* of three or four
fingers length ‖.

Thefe birds run very fwiftly ¶, and fly in
flocks: they are migratory in France, and hardly
ftop in the interior provinces; but they refide
in the maritime diftricts, as in Poitou **, Aunis,
and in Brittany along the Loire, where they
breed ††. It is affirmed, that in England
they inhabit the coafts only in winter, and that
in fummer they retire to neftle in the upland
country ‡‡. In Germany they arrive in rainy
weather when the wind is in a certain quarter;
for the different names there applied to them
allude to torrents, or tempefts. They are feen
alfo in Silefia about autumn ‖‖, and they advance

* Gefner.
† Albin.
‡ Willughby.
§ *Idem.*
‖ *Idem.*
¶ Hence probably Hefychius has erroneoufly applied the name
trochilus to the Curlew, which belongs to the gold-crefted wren.
Clearchus indeed mentions a *trochilus*, which muft be either the
courier, or fome of the fmall dunlins or collared plovers, which
frequent the fhores and run with fpeed.
** In Poitou thoufands are feen entirely gray. *Salerne.*
†† *Idem.*
‡‡ Britifh Zoology, and Borlafe's *Nat. Hift. of Cornwall.*
‖‖ Schwenckfeld.

in fummer as far as the Baltic fea *, and the gulph of Bothnia †. They are found too in Italy and in Greece ; and it appears that their migrations extend beyond the Mediterranean, for they pafs Malta twice a year, in the fpring and autumn ‡. Voyagers have difcovered Curlews in almoſt every part of the world § ; and though their accounts refer for the moſt part to different foreign branches of this family, it appears the European kind occurs at Senegal ‖, and in Madagaſcar, fince the bird reprefented,

* Klein.

† *Fauna Suecica.* Brunnich. *Ornith. Boreal.*

‡ Obfervation communicated by the Commander Defmazy.

§ Curlews are found in New Holland and in New Zealand. *Cook.*—Numerous in the falt-marſhes of Tinian. *Anfon.*—In Chili. *Frezier.*—In an excurfion on Statenland, we took new fpecies of birds, among others a handfome gray curlew ; its neck was yellowiſh ; it was one of the moſt beautiful birds we had ever feen. *Forfter.*—In the iſle of May (one of the Cape de Verd Iſlands) we found curlews. *Roberts.*—The country of Napal breeds different forts of birds . . . great numbers of ducks . . . Others are very much like our curlews, their fleſh hard, but good to eat, *Dampier.*—In the bay of Campeachy there are ducks, curlews, pelicans, &c. *Idem.*—There are two forts of curlews, that differ in bulk as well as in colour ; the largeſt are equal to turkey-cocks ; (this feems exaggerated) their legs are long, and their bill hooked ; they are of a dull colour ; their wings are mixed with black and white ; their fleſh is black, but very good and wholefome. The Engliſh call them *double curlews*, becaufe they are twice as large as the biggeſt of the others. The little curlews are of a dull brown, their legs and their bill are the fame with thofe of the preceding ; they are more efteemed than the others, becaufe their fleſh is more delicate. *Idem.*

‖ There are many water fowl in the marſhes of Senegal, fuch as curlews, woodcocks, teals. *Adanfon*, p. 138.

N° 198,

N° 198, *Planches Enluminées*, is fo like our Curlew, that it may be regarded as of the fame fpecies, differing only by the greater length of its
bill and the diftinctnefs of its colours.   Sometimes White Curlews are to be met with *, but
they are only individual varieties or accidental
degradations.

* Salerne.

[A] Specific character of the Curlew, *Scolopax-Arquata.* " Its
bill is arched; its feet blueifh; its wings black, with fnowy
fpots."   It lays four eggs of a pale olive, marked with dufkifh
fpots. It weighs between twenty-two and thirty-feven ounces,

The WHIMBREL.

Le Corlieu, ou Petit Courlis. *Buff.*

SECOND SPECIES.

Scolopax-Phæopus. Linn. and Gmel.
Numenius Minor. Briff.
Arquata Minor. Ray and Will.
 * *Numenius-Phæopus.* Lath.

T<small>HE</small> Whimbrel is one half lefs than the common curlew, which it refembles in its form, in the ground of its colours, and even in their diftribution; it has alfo the fame habits and mode of life. Yet they are two diftinct fpecies; for, befides their great inequality of bulk, they never affociate together. The Whimbrel feems in particular to be attached to England, where, according to the authors of the Britifh Zoology, it is more frequent than the curlews †. On the contrary, it is very rare in the provinces of France, and is probably not more common in Italy; for Aldrovandus gives

* In Italian *Tarangolo*, or *Taraniolo*: in Danifh *Mellem-Spove*: in Norwegian *Smaae Spue*.

† This is a miftake; Mr. Pennant fays directly the reverfe " The Whimbrel is much lefs frequent on our fhores than the " curlew."—T.

but

THE WHIMBREL.

but a confufed account of it from Gefner, and copies the miftake of that naturalift, who introduces the Whimbrel twice among the aquatic birds, under the different names *phæopus* and *gallinula*. Willughby firft noticed this overfight of Gefner's. The *little ibis* defcribed by Edwards, is undoubtedly a Whimbrel, only its plumage is altered by moulting, as that naturalift remarks *,

* Mr. Edwards' *little ibis* is certainly a Whimbrel, only it was in moult. The bird *torea* of the Society Ifles, called in Cook's Voyage *a little curlew*, feems not to belong to that family, fince it is faid to be found *about fhips*.

[A] Specific character of the Whimbrel, *Scolopax-Phæopus:* " Its bill is arched, its feet blueifh, brown rhomboidal fpots on its back." Its weight exceeds not twelve ounces.

The GREEN, or ITALIAN CURLEW.

THIRD SPECIES.

Tantalus-Falcinellus. Linn. and Gmel.
Numenius-Viridis. Briff.
Numenius Subaquilus. Klein.
Falcinellus. Gefn. Johnft. Will.
The Scythe Bill. Ray.
The Bay Ibis. Penn. and Lath.

THIS is nearly as large as a heron, according
to Aldrovandus, and among the Italians it
has fometimes that name. The appellation
falcinellus, which that naturalift and Gefner
feem to have beftowed only on this fpecies,
might be extended to all the other curlews,
which have their bills equally hook-fhaped.
In the prefent, the head, the neck, the fore
part of the body, and the fides of the back, are
of a fine deep chefnut; the upper fide of the
back, of the wings, and of the tail, are green
gloffed with gold or bronze, according to pofition with regard to the light; the bill is blackifh, as well as the feet and the naked part of the
leg. Gefner defcribes only a yellow bird, which
had not attained its fize or its colours. This

curlew,

curlew, which is common in Italy, occurs too in Germany *; and the Danube curlew of Marfigli, cited by Briffon, is apparently a variety of the fame fpecies.

* According to Gefner, it is called in German *Welfcher-Vogel*, (Italian bird) ; *Sichler* (fickler) ; *Sagifer* (fawyer.)

[A] Specific character of the *Tantalus-Falcinellus* : " Its face is black, its feet blue, its wings and tail violet, its body chefnut."

The BROWN CURLEW.

FOURTH SPECIES.

Tantalus Manillensis. Gmel.
The Manilla Ibis. Lath.

SONNERAT found this Curlew in the isle of Luçon, one of the Philippines. It is as large as the common European curlew: all its plumage is rufous brown; its eyes are encircled with a greenish skin; its iris flame-coloured; its bill greenish; and its feet of a lacker-red.

The SPOTTED CURLEW.

FIFTH SPECIES.

Scolopax Luzonienſis. Gmel.
The Luzonian Curlew. Lath.

THIS Curlew alſo is found in the iſle of
Luçon. It too reſembles the European
kind, only is one-third ſmaller. It is diſtinguiſhed beſides, becauſe the crown of its head is
black, and its colours differently diſtributed;
they are ſcattered on the back in ſtreaks on the
edge of the feathers, and on the belly in waves
or tranſverſe breaks.

The BALD CURLEW.

LE COURLIS A TETE NUE. *Buff.*

SIXTH SPECIES.

Tantalus Calvus. Gmel.
The Bald Ibis. Lath.

THIS fpecies of Curlew is new and very fingular : its whole head is naked, and on the top is a fort of roll five lines thick, flattened back, and covered by a very red and thin fkin, immediately under which we perceive a bony protuberance ; the bill is of the fame red with this crown ; the top of the neck and the fore part of the throat are alfo bare of feathers ; and the fkin is, no doubt, vermilion in the living fubject, but was livid in the dried fpecimen which we defcribe, and which was brought from the Cape of Good Hope by M. de la Ferté. It has entirely the form of the European curlews, only ftronger and thicker ; the ground of its plumage is black, and on the feathers of the wings there is a varying green and purple glofs ; the fmall coverts of the wings are of a deep purple violet, but lighter on the back,

back, the neck, and the under fide of the body;
the feet and the naked part of the leg, for the
fpace of an inch, are red like the bill, which is
four inches and nine lines long. This Curlew,
meafured from the point of the bill to the extremity of the tail, is two feet and an inch, and
in its natural attitude it is a foot and an half
tall.

The CRESTED CURLEW.

SEVENTH SPECIES.

Tantalus Criftatus. Gmel.
The Crefted Ibis. Lath.

THE creft diftinguifhes this Curlew from all
the reft, in which the head is more or
lefs fmooth, or covered with very fhort little
feathers; this, on the contrary, has a fine tuft
of long feathers, partly white and partly green,
which fall back; the fore-fide of the head, and
the compafs of the top of the neck, are green;
the reft of the neck, the back, and the fore-part
of the body, are of a fine chefnut rufous; the
wings are white; the bill and feet are yellowifh;
a broad portion of naked fkin furrounds the
eyes; the neck, which is well clothed with
feathers, is not fo long or flender as in the
other curlews. This beautiful bird is found in
Madagafcar.

These feven fpecies of Curlews belong all to
the ancient continent; there are eight which
inhabit the new.

CURLEWS of the New Continent.

The RED CURLEW*.

FIRST SPECIES.

Tantalus Ruber. Linn. and Gmel.
Numenius Brafilienfis Coccineus. Briff.
Guara Brafilienfibus. Marcg. Will. Johnft. &c.
Avis Porphyrio Amboinenfis. Seba.
Numenius Ruber. Klein.
The Scarlet Ibis. Lath. and Penn.

THE low flimy grounds contiguous to the fea, and the great rivers of South America, are inhabited by many fpecies of Curlews: the moft beautiful of thefe, and the moft common in Guyana, is the prefent: all its plumage is fcarlet, except the tip of the firft quills of the wing, which is black; the feet, the naked part of the legs, and the bill, are red or reddifh †, and alfo the bare fkin that covers the fore part of

---

* Buffon and Catefby.

† This colour of the bill may vary. Marcgrave fays, that it is *cinereous-white :* Clufius, that it is *ochry-yellow.*

the head, from the origin of the bill to beyond
the eyes. This Curlew is large, but not fo
thick as the European; its legs are taller, and
its bill longer and ftronger, and much thicker
near the head. The female has its plumage of
a fainter red than the male *, and neither of
them acquires that beautiful colour till the proper age; for at firft they are covered with a
blackifh down †, then cinereous, and afterwards
white, when they begin to fly ‡, fo that the
fine red is introduced by fucceffive gradations,
does not appear before the fecond or third year,
and turns brighter as the bird grows older.

Thefe birds keep together in flocks, whether
they fly or perch on trees, where their number
and their flame-coloured plumage render them
confpicuous objects §. Their motion through
the air is fteady and even rapid, but is performed only in the morning and evening; during the heat of the day they enter the creeks,
and enjoy the cool fhade of the mangroves; at
three or four o'clock they return to the mud,
which they again quit to pafs the night under
the branches and foliage. Seldom one of thefe
Curlews is feen alone, or if one fhould happen

* Catefby.
† Marcgrave.
‡ De Laët.
§ " The guaras fly in flocks, and their fcarlet plumage forms
" a very beautiful fpectacle in the beams of the fun." *Hift. Gen.
des Voy.* tom. xiv. p. 304.

to

to ftray from the flock, it haftens to join its companions. But thefe focieties are diftinguifh-ed by their age, the old birds keeping feparate from the young. Their hatches begin in Ja-nuary and end in May; they lay their eggs, which are greenifh, in the large plants that grow under the mangroves, or amidft the bram-bles, on fome fticks collected. The young ones may eafily be caught by the hand, even when the mother leads them out to fearch for infects and fmall crabs, which are their principal food: they are not wild, and they foon become recon-ciled to the domeftic ftate. " I reared one," fays M. de la Borde, " which I have kept up-
" wards of two years; it fed out of my hand
" very familiarly, and never miffed the time of
" dinner and fupper: it ate bread, flefh either
" raw, dreffed, or falted, fifh, every thing in fhort
" was acceptable; it fhowed however a pre-
" ference to fowls' and fifhes' guts, and with that
" view it frequently paid a vifit to the kitchen.
" At other times it was conftantly employed
" feeking earth-worms, either round the houfe
" or in the garden befide the negro who was at
" work on the ground. In the evening it re-
" tired of its own accord into a hen-houfe, where
" it repofed with an hundred fowls: it roofted
" on the higheft bar, and with violent ftrokes of
" its bill drove off all the hens that had occu-
" pied its place; and often during the night it
" took pleafure in annoying its fellow-lodgers.

" It

" It was rouzed early in the morning, and be-
" gan by making three or four circuits round
" the houfe; fometimes it went to the fea-
" fhore, but did not ftop there. I never heard
" it utter any cry except a little croaking, which
" feemed to be an expreffion of fear at the fight
" of a dog or other animal. It had a great anti-
" pathy to cats, but did not fear them; ran fierce
" and undaunted upon them. It was killed near
" the houfe in a bog by a fportfman, who took
" it for a wild curlew."

This account given by M. de la Borde, correfponds with that of Laët; who adds, that he
has feen fome of thefe birds copulate and breed
in the ftate of domeftication. We prefume,
therefore, that it would be equally eafy and agreeable to rear and propagate this beautiful fpecies,
which would be an ornament to our court-
yards *, and add perhaps to the pleafures of
the table; for its flefh, which is already tolerably
pleafant, might be improved, and might lofe its
flight marfhy tafte †: befides, living on the
offals and garbage of the kitchen, it would coft
little for maintenance.—We know not whether,

* At the time that I wrote this, there was a Red Curlew living
in the menagerie of his Royal Highnefs the Prince of Conde, at
Chantilly.

† " It is eaten in ragouts, and makes tolerable fauce; but it
" muft be previoufly half-roafted to difcharge part of its oil,
" which has a brackifh tafte." *Note given by a colonift of Cayenne.*—
" The flefh of the curlew is a difh much efteemed." *Effay on the
Natural Hiftory of Guiana,* p. 172.

as

as Marcgrave fays, this Curlew foaks previoufly in water whatever it eats.

In the ftate of nature, thefe birds live on fifh, fhell-fifh, and infects, which they find in the flime at ebb-tide. They never go very far from the fea-coaft, nor advance up the rivers to a confiderable diftance from their mouths. They refide through the whole year in the fame dif-trict, only fhifting from one part to another. The fpecies is however diffufed through moft of the hot countries of America *. It is found at the mouths of the *Rio-Janeiro* †, of the *Ma-ragnon*, &c. in the Bahama Iflands ‡, and in the Antilles §. The Indians of Brazil, who are fond of decking themfelves with their beautiful feathers, call thefe Curlews by the name of *guara* ||. The appellation *flammant*, which they receive in Cayenne, refers to the flame-colour of their plumage; and the colonifts have very im-properly beftowed the fame term on all the cur-lews. With equal inaccuracy the voyager Cauche confounds with it his violet Madagafcar Curlew ¶.

* Catefby.
† Marcgrave.
‡ Catefby.
§ Sloane.
|| Barrere.
¶ " The herons of this country (Madagafcar) have large thick
" bills, which bend gradually downwards after the fafhion of a
" Polifh cutlafs; their feathers are violet; the wings terminate
" with the tail; their thighs, as far as the knot of the leg, are co-
" vered with little feathers; their legs long and wafhed with gray:

" the

" the chicken is black, and as it grows it turns cinereous, then
" white, then red, and at laſt columbine, or light violet : it lives on
" fiſh.   There are ſimilar birds in Brazil called *guara* ; the figure
" occurs in Marcgravius." *Voyage à Madagaſcar & au Breſil,* par
Franc. Cauche, *Paris* 1651. p. 133.

[A] Specific character of the *Tantalus Ruber* : " Its bill, its
" face, and its feet, are red ; its body blood-coloured ; the tips of
" its wings black."

The WHITE CURLEW*.

SECOND SPECIES.

Tantalus Albus. Gmel.
Scolopax Alba. Linn.
Numenius Albus. Klein.
Numenius Brasiliensis Candidus. Briss.
The White Ibis. Penn. and Lath.

WE might reckon this a red curlew having its first colour; but Catesby, who knew both, conceives it to be a different species. It is larger; its feet, its bill, its orbits, and the foreside of its head, are of a pale red; all the plumage is white, except the four first quills of the wing, which, at their extremity, are of a dull green. Great numbers of these birds arrive in Carolina about the middle of September, which is the rainy season: they frequent the low marshy grounds, where they remain about six weeks, and then disappear; retiring probably to the south, to breed in a warmer climate. Catesby says, that he found clusters of eggs in many females shortly before their departure from Carolina. They differ not from the males in regard to colours; both of them have their flesh and fat yellow, like the pheasant.

* Buffon and Catesby.

D 4

The RED-FRONTED BROWN CURLEW.

THIRD SPECIES.

Tantalus Fuscus. Gmel.
Scolopax Fusca. Linn.
Numenius Brasiliensis Fuscus. Briff. and Klein.
Arquata Cineree. Barrere.
The Brown Curlew. Catesby.
The Brown Ibis. Penn. and Lath.

THESE Brown Curlews arrive in Carolina with the white curlews, and intermingled with their flocks. They are of the same size, but fewer, *" there being twenty white curlews,"* says Catesby, *" to one brown."* They are entirely brown on the back, the wings, and the tail; brown-gray on the head and the neck; and all white on the rump and the belly: the fore part of the head is bald, and covered with a pale red skin; and the bill and feet are also of that colour. Like the white curlews, they have yellow flesh and fat. Both species arrive and depart together; they pass in winter from Carolina to the more southern climates, such as Guyana, where they are termed the *gray flammants.*

The WOOD CURLEW.

LE COURLIS DES BOIS. *Buff.*

FOURTH SPECIES.

Tantalus Cayanenfis. Gmel.
The Cayenne Ibis. Lath.

THIS fpecies, which the fettlers at Cayenne denominate the *wood flammant*, lives in the forefts befide the brooks and rivers, and far from the fea-coaft, which the other curlews feldom ever leave. Its habits too are different; it never goes in flocks, but only in company with its female. It fifhes, fitting on wood that floats in the water. It is not larger than the green curlew of Europe, but its cry is much ftronger. Over its whole plumage is fpread a very deep green tint, on a dull brown ground, which at a diftance appears black, and viewed near exhibits rich blueifh or greenifh reflections: the wings and the top of the neck have the colour and luftre of polifhed fteel; on the back are bronze reflections, and on the belly and the lower part of the neck a purple glofs: the cheeks are bare of feathers. Briffon takes no notice of this fpecies, though Barrère has mentioned it twice under the appellations of *arquata viridis fylvatica,* and *flammant des bois.*

[A] Specific charaɛ̸ter of the *Tantalus Cayanenfis :* " Its face is " dull reddifh, its bill obfcure; its body black, with a green glofs."

4

The G U A R O N A.

FIFTH SPECIES.

Scolopax-Guarauna. Linn. and Gmel.
Numenius Americanus Fuscus. Briff.
Numenius-Guarauna. Lath. Ind.
Guarauna. Pifon. Marcg. Johnft. Ray. and Will.
The Brafilian Whimbrel. Lath. Syn.

GUARA, we have feen, is the Brazilian name of the red curlew. *Guarana* or *Guarona* is beftowed on this fpecies, whofe plumage is chefnut-brown, with green reflections on the rump, on the fhoulders, and on the outer edge of the quills of the wing; the head and neck are variegated with fmall longitudinal whitifh lines on a brown ground. This bird is two feet long from the bill to the nails * : it bears a great refemblance to the green curlew of Europe, and appears to be the reprefentative of that fpecies in the new world. Its flefh is tolerably good, according to Marcgrave, who fays that he often ate of it. It occurs both in Brazil and in Guyana.

* Marcgrave fays, that it is of the bulk of the *iacu*; but the yacou is fcarcely fo large as an ordinary hen, a fize which exactly correfponds to a curlew.

[A] Specific character of the *Scolopax-Guarauna:* " Its bill is " arched and yellowifh; its feet brown; its head brown ftriped " with white."

The A C A L O T.

SIXTH SPECIES.

Tantalus Mexicanus. Gmel.
Numenius Mexicanus Varius. Briff.
Corvus Aquaticus. Nieremb. Fernand. Will. &c.
The Mexican Ibis. Lath.

WE abridge the name *acacalotl,* beftowed on this curlew in Mexico, into *Acalot :* it is indigenous in that country; and, like moft of the reft, its front is bald and covered with a reddifh fkin : its bill is blue; the neck and back of the head clothed with feathers, which are brown intermixed with white and green ; the wings fhine with green and purple reflections. And thefe characters have probably induced Briffon to denominate it the *variegated curlew* ; but it is eafy to fee, from the appellation of *water raven* given by Fernandez and Nieremberg, that thefe colours are laid on a dark ground approaching to black. Adanfon, remarking that this bird differs from the European curlews in having its front bald, ranges it, on account of that property, with the ibis, the *guara,* and the *curicaca,* of which he makes a diftinct genus. But the character by which he difcriminates it from the

curlews

curlews appears infufficient; fince it has in other refpects a fimilar form, and that difference is introduced by fucceffive gradations, infomuch that fome fpecies, the green curlews for inftance, have only a bare fpace round the eyes, while others, fuch as the prefent Acalot, are naked on a great part of the front. We have feparated the *curicaca* from the curlews, on account of its magnitude and fome other effential differences, particularly the fhape of its bill.—We do not underftand why this learned naturalift claffed thefe birds with the lapwings *.

* See Supplement to the Encyclopedie, article *Acacalotl.*

The SHORE MATUITUI.

LE MATUITUI DES RIVAGES. *Buff.*

SEVENTH SPECIES.

Tantalus Grifeus. Gmel.
Numenius Americanus Minor. Briff.
Matuitui. Pifon. Marcg. Will. Johnft. &c.

IF we were better acquainted with this bird,
we fhould perhaps feparate it, as well as the
curicaca, from the curlews; fince Marcgrave and
Pifon fay that it is like the curicaca, though on
a fmaller fcale, which is disjoined from the curlews, both by its bulk and the character of its
bill; and till that character be known we cannot
affign its rank. We may obferve, however, that
the appellation of *little curlew* given by Briffon
is improper, for it is nearly as large as a hen,
and therefore of the firft magnitude in the genus of curlews.——This Shore Matuitui differs
from the other little matuitui mentioned by
Marcgrave in another place, which is hardly
larger than a lark, and appears to be a little
ringed plover.

The GREAT CURLEW of CAYENNE.

EIGHTH SPECIES.

Tantalus Albicollis. Gmel.
The White-necked Ibis. Lath.

IT is larger than the European curlew, and
seems to be the greatest of all the curlews.
The whole of its upper surface, the great quills
of its wings, and the fore side of its body, is
brown, waved with gray and glossed with green;
the neck is rusty white, and the great coverts of
the wing are white. This description suffices
to distinguish it from the rest of the curlews.

THE LAPWING.

The LAPWING.

Le Vanneau. *Buff.*

FIRST SPECIES.

Tringa-Vanellus. Linn. and Gmel.
Capella. Gesner.
Vanellus. Aldrov. Ray, Will. Johnst. Sibb, &c.
Gavia Vulgaris. Klein.
The Lapwing, Bastard Plover, or Pewit. Alb. Will. Penn. and Lath.

THE * appellation of this bird, in modern Latin, in French, and in English, alludes to the incessant flapping of its wings. The Greeks, besides giving it other names expressive of its cry, denominated it *the wild peacock* (Ταως αγριος), on account of its crest and its elegant colours: yet this crest is very different from

---

* The Greeks applied to this bird the name of goat, Αιξ, and Αιγα, on account of its cry: in modern Latin it is for the same reason termed *Capella*; the term *Vanellus*, from *Vannus*, a fan, was given it because of the frequent and noisy flapping of its wings: and hence too the French name *Vanneau*, and the English Lapwing. In German it is called *Kwyit* and *Himmel-Geiss (sky-goat)*: in Swiss *Gyfitz, Gywitz, Blaw Gruner Gyfitz*: in Dutch *Kwidt*: in Portuguese *Byde*: in Polish *Czayka Kozielek*: in Swedish *Wipa, Kowipa Blæcka*: in Danish *Vibe Kivit*: in Turkish *Gulguruk*: in Italian *Paonzello*, or *Pavonzino*, (i. e. little peacock): in many of the French provinces it is termed *dix-huit, pivite, kivite*: in some parts of England it has the appellation *pewit*: and all these names, and many others also, have nearly the same sound, and are evident imitations of the bird's cry, *pēw-ēēt*.

that

that of the peacock, it confisting only of some long unwebbed and very slender feathers; and of its plumage, the under side is white, the upper of a dark cast, and it is only when held close to the eye that we can perceive the brilliant gold reflections. In some parts of France, the Lapwing has the denomination of *dix-huit* (eighteen) because these two syllables, pronounced faintly, express, with tolerable accuracy, its cry. which many languages have endeavoured to denote by imitative sounds *. In rising up it vents one or two screams, which it often repeats at intervals as it flies, even during the night † : its wings are powerful, and much exercised; for in the air it long maintains its flight, and rises to a great height, and on the ground it springs and bounds, and skims from spot to spot.

The Lapwing is joyous, and perpetually in motion; it sports and frolics a thousand ways in the air; it assumes, at times, every imaginable posture, its belly sometimes even turned upwards or sidewise, and its wings expanded perpendicularly; and no bird wheels and flickers so nimbly.

The Lapwings arrive in our meadows in great flocks about the beginning of March, or even as early as the end of February, after the first open weather, when the wind is southerly. At this

* *Gyfytz, Giwitz, Kiwitz, Czieik,* &c.
† It imitates the tremulous voice of a goat, while it flies in the night-time. *Rzacynski.*

feafon they alight in the fields of green corn *,
or in the morning cover the low marfhy grounds
in fearch of worms, which they dextroufly draw
from their holes : when the bird meets with one
of thofe little clufters of pellets, or rolls of earth,
which are thrown out by the worm's perfora-
tions, it firft gently removes the mould from the
mouth of the hole, ftrikes the ground at the fide
with its foot, and fteadily and attentively waits the
iffue; the reptile, alarmed by the fhock, emerges
from its retreat, and is inftantly feized †. In
the evening the Lapwings purfue a different plan;
they run along the grafs, and feel under their
feet the worms, which now come forth, invited
by the coolnefs of the air: thus they obtain a
plentiful meal, and afterwards they wafh their
bill and feet in the fmall pools or rivulets.

These birds are difficult to be approached,
and feem to defcry the fowler at a great diftance: we can gain nearer them in a ftrong
wind, for then they fly with difficulty. When
they are congregated and ready to rife together,
they all flap their wings with an equal motion;
and as they keep clofe to each other, and their

---

* Belon, Nat. des Oifeaux, *liv.* iv. 17.

† " To afcertain this circumftance," fays M. Baillon, " I employed
" the fame ftratagem : in a field of green corn, and in the garden, I
" beat the earth for a fhort time, and I faw the worms coming out ;
" I preffed down a ftake, which I then turned in all directions to
" fhake the foil: this method, which is faid to be ufed by the cur-
" lews, fucceeds ftill quicker; the worms crawled out in crowds,
" even at the diftance of a fathom from the ftake."

under fide is white, the ground, which was darkened by their numbers, appears at once white. But this great fociety, which thefe birds form on their arrival, diffolves when the vernal warmth invites to love, and in two or three days they difperfe. The fignal is given by battles between the males ; the females feem to avoid the contentions, and firft abandon the flock, as if unconcerned in the quarrels : but, in fact, they draw off the combatants to form a fweeter and more intimate union, which lafts three months.

The hatch is conducted in April; it confifts of three or four oblong eggs, of a dull green, much fpotted with black : thefe are dropped in the marfhes, on the little heads or clods of earth raifed above the furface of the plain ; a precaution which feems neceffary to guard againft the accidental fwelling of the water, but which, however, leaves the neft expofed. To make a fite for it, they are contented with cropping, clofe to the furface, a little round fpace in the grafs, which foon withers about it, from the heat of incubation; and if we find the grafs frefh and verdant, we may infer that the eggs have not been covered. It is faid, that thefe eggs are good to eat, and in many provinces great quantities are gathered for market. But is it not an incroachment on the rights of nature, an invafion on her property, to deftroy thus the tender germs of fpecies which we cannot multiply?

The

6

The eggs of domeftic poultry are in a manner our own creation, but thofe of independent birds belong only to the common mother of all.

The incubation of the Lapwing, as in moft other birds, lafts twenty days: the female fits affiduoufly; if any thing alarms it, and drives it from its neft, it runs a little way, cowering through the grafs, and does not rife till at a good diftance from its eggs, that it may not betray the fpot. The old hens, whofe nefts have been robbed, will not again breed expofed in the marfhes; they retire among the growing corn, and there in tranquillity make their fecond hatch: the young ones, lefs experienced, are not deterred by their lofs, and they rifk their neft a fecond, or even a third time in the fame place; but thefe after-layings never exceed one or two eggs.

The young Lapwings, two or three days after being hatched, run among the grafs and follow their parents: thefe from folicitude often betray the little family, and difcover the retreat, as they flutter backwards and forwards over the fowler's head with cries of inquietude, which are augmented as he approaches the fpot where the brood had fquatted on the firft alarm. When pufhed to extremity they betake themfelves to running, and it is difficult to catch them without the affiftance of a dog, for they are as alert as partridges. At this age they are covered with

E 2 a blackifh

a blackifh down, fhaded under with long white hairs; but in July they drop this garb, and acquire their beautiful plumage.

The great affociation now begins to be renewed: all the Lapwings of the fame marfh, young or old, affemble; thofe of the adjacent marfhes join them, and in a fhort time, a body of five or fix hundred are collected. They hover in the air, faunter in the meadows, and, after rain, they difperfe among the plowed fields.

Thefe birds are reckoned inconftant, and indeed they feldom remain above twenty-four hours in the fame tract: but this volatility is occafioned by the fcantinefs of food; if the worms of a certain haunt be confumed in one day, the flock muft remove on the following. In the month of October the Lapwings are very fat, and this is the time when they live in greateft abundance; becaufe in this wet feafon the worms fwarm on the furface; but the cold winds which blow about the end of the month conftrain them to retire into the earth, and thus oblige the Lapwings to pafs into another climate. This is the general caufe of migration in the vermivorous birds. On the approach of winter, they advance towards the fouth, where the rains are only begun, and, for a like reafon, they return in the fpring; the exceffive heat and drynefs of the fummer in thofe latitudes having the fame effect as great cold in confining the worms in

the

the ground *. And that the time of migration
is the fame throughout the whole of our hemi-
fphere, is evinced by this circumftance, that at
Kamtfchatka October is denominated *the month
of Lapwings* †; which, as in our latitudes, is the
time of their departure.

Belon fays, that the Lapwing is *known over the
whole earth*; and the fpecies is indeed widely difperfed. We have juft mentioned their being
found in the eaftern extremity of Afia; they are
met with alfo in the interior parts of that vaft
region ‡, and they are feen in the whole of Europe. In the end of winter, thoufands of them
appear in our provinces of Brie and Champagne §, and great numbers are caught. Nets

---

* M. Baillon, to whom we are indebted for the beft details in this
hiftory of the Lapwing, confirms our idea with refpect to the caufe of
the return of the Lapwing from fouth to north, by an obfervation
which he made himfelf in the Antilles. " The ground," fays he,
" is, during fix months of the year, extremely hard and parched in
" the Antilles; not a fingle drop of rain falls in the whole of that
" time; I have feen cracks in the valleys four inches broad, and feve-
" ral feet deep; no worm can then live at the furface: accordingly,
" in the dry feafon, no vermivorous bird is obferved in thefe iflands;
" but on the firft days after the rains fet in, they arrive in fwarms,
" and come, I fuppofe, from the low deluged lands on the eaftern
" fhores of Florida, from the Bahamas, and a multitude of other
" iflands, lying north or north-weft from the Antilles: all thefe wet
" places are the cradle of the water-fowl of thefe iflands, and per-
" haps of a part of the great continent of America."

† *Pikis koatch*; *pikis* is the name of the bird. *Gmelin.*

‡ " The Lapwings are very numerous in Perfia." *Lettres Edifiantes, trentieme Recueil*, p. 317.

§ " In this province, and particularly in the canton of Baffigny,
" they are hunted at night with flambeaux; the light wakes them,
" and, it is faid, attracts them." *Note communicated by M. Petitjean.*

are

are spread in a meadow, and a few stakes and one
or two live Lapwings set in the middle space to
entice the birds; or the fowler, concealed in his
lodge, imitates their cry with a call made of fine
bark; and the whole flock, thus betrayed, alight
and are ensnared.   Olina mentions the course of
November as the time of the greatest captures;
and from his account it appears, that in Italy
the Lapwings remain congregated the whole
winter *.

The flesh of the Lapwing is held in considerable estimation †; yet those who have drawn the
nice line of pious abstinence have, by way of
favour, admitted it into the diet of mortification.
This bird has a very muscular stomach, lined
with an inadhesive membrane, covered by the
liver, and containing, as usual, a few small pebbles; the intestinal tube is about two feet long:
it has two *cæca* directed forwards, each more
than two inches long; a gall-bladder adhering to
the liver and the *duodenum:* the liver is large,
and divided into two lobes ‡; the *œsophagus*
about six inches long, dilated into a bag before
its insertion; the palate is rough with small
fleshy points, which lie backwards; the tongue

* M. Hebert assures us, that a few remain in Brie till the depth
of winter.

† It is much valued in some of the provinces: in Lorraine there
is an old proverb, *Qui n'a pas mange de Vanneau, ne sait pas ce que
gibier vaut* (He that has not eaten Lapwing, knows not what game
is worth).

‡ Willughby.

is

is narrow, rounded at the tip, and ten lines in length. Willughby obferves, that the ears are placed higher in the Lapwing than in other birds.

There is no diftinction, in point of fize, between the male and the female, but, in the colours of the plumage, fome differences occur, though Aldrovandus fays, that he did not perceive any. The tints of the female are in general more dilute, and the black parts mixed with gray: its creft is alfo fmaller than that of the male, whofe head feems to be rather larger and rounder. In both the feathers are thick and well clothed with down, which is black near the body: the under fide and the verge of the wings, near the fhoulders, are white, and alfo the belly, the two outer feathers of the tail, and the firft half of the reft: there is a white point on each fide of the bill, and a ftreak of the fame colour on the eye: all the reft of the plumage is of a black ground, but enriched by fine reflections of a metallic luftre, changing into green and gold-red, particularly on the head and the wings: the black on the throat and the forepart of the neck is fpotted with white, but on the breaft it forms alone a broad round fpace, and, like the black of the wings, it is gloffed with bronze green: the coverts of the tail are rufous:—but as the plumage frequently varies fomewhat in different individuals, it will be unneceffary to be more particular in the defcrip-

E 4   tion;

tion; I ſhall only obſerve, that the tuft is not
inſerted in the front, but in the back of the
head, which is more graceful; it conſiſts of five
or ſix delicate threads, of a jet black the two
upper ones cover the reſt, and are much longer.
The bill is black, pretty ſhort and ſmall, not exceeding twelve or thirteen lines, inflated near the
point: the feet are tall and ſlender, and of a
brown red, as well as the lower part of the legs,
which is naked for the ſpace of ſeven or eight
lines, the outer and middle toes are joined at
their origin by a ſmall membrane; the hind one
is very ſhort, and does not reach the ground: the
tail does not extend beyond the wings when they
are cloſed: the total length of the bird is eleven or twelve inches, and its bulk is nearly that
of a common pigeon.

Lapwings may be kept in the domeſtic ſtate;
" they ſhould be fed," ſays Olina, " with ox-
" heart minced in ſhreds." Sometimes they are
let into the gardens, where they are uſeful in de-
deſtroying inſects *: they remain contented, and
never ſeek to eſcape; but, as Klein remarks, the
facility in the domeſtication of this bird, proceeds
rather from its ſtupidity than its ſenſibility; and
that obſerver aſſerts, that the demeanour and

* " I have often had Lapwings in my garden; I have ſtudied them
" much; they were reſtleſs, like quails, at the time of migration,
" and ſcreamed immoderately for ſeveral days. I accuſtomed them
" to live on bread and raw fleſh in winter; I kept them in a cellar,
" but they grew very lean." *Note communicated by M. Baillon.*

physiognomy

phyfiognomy of both the Lapwings and plovers
fhew that their inftincts are obtufe.

Gefner fpeaks of white lapwings, and of
brown fpotted lapwings without the tuft: but
of the firft his account is not fufficiently precife
for us to judge whether they are not merely
accidental varieties; and with regard to the fecond, he feems to miftake plovers for Lapwings, for he elfewhere confeffes that he was
little acquainted with the plover, which is extremely rare in Switzerland, while Lapwings
are very frequent; and there is even a fpecies
called the Swifs Lapwing.

[A] Specific character of the Lapwing, *Tringa-Vanellus:* "Its
" feet are red, its creft hanging, its breaft black." It remains in
Great Britain the whole year, though it often fhifts its haunts: its
eggs are fold as great delicacies, by the London poulterers, at three
fhillings a dozen.

The SWISS LAPWING.

SECOND SPECIES.

Tringa Helvetica. Linn. and Gmel.
Vanellus Helveticus. Briff.
Charadrius Hypomelus. Pallas.
The Swifs Sandpiper. Penn. and Lath.

T H I S is nearly as large as the common lapwing; all the upper fide of the body is variegated tranfverfely with waves of white and
brown; the fore-part of the body is black or
blackifh; the belly is white; the great quills of
the wings are black, and the tail is croffed with
bars like the back; it might therefore derive its
denomination of Swifs Lapwing from its parti-
coloured garb; which is perhaps as plaufible a
fuppofition as that it received this name from
its greater frequency in Switzerland *.

Briffon makes the *ginochiella* of Aldrovandus a
third fpecies, under the appellation of the *greater
lapwing* †, which little belongs to that bird,

---

* There is a very cogent reafon for doubting whether this
bird be found at all in that country, fince fo intelligent an obferver
as Gefner makes no mention of it.

† Tringa Bononienfis. *Linn.* and *Gmel.*

Specific character: " Its feet ochry, its head and neck bay, its
" body black above and white below; its throat and breaft marked
" with ferruginous fpots."

fince

fince Aldrovandus's figure, which he fays is the natural fize, reprefents it as fmaller than a common lapwing. But it is difficult to decide on the reality of a fpecies from the fight of an imperfect figure; particularly as, unlefs the bill and feet be badly delineated, it cannot be a lapwing; we might rather clafs it with the *great plover*, or *land curlew*, of which we fhall fpeak at the clofe of the article of the plovers, if the difference of its fize had not oppofed this arrangement. Aldrovandus, in the fhort account which he fubjoins to his figure, fays, that its bill has a fharp point, a property which belongs equally to the plover and to the lapwing: fo that we fhall content ourfelves with juft mentioning this bird, without venturing to decide its fpecies.

[A] Specific character of the *Tringa Helvetica :* " Its bill and " feet are black, its under fide black, its vent white, its tail-quills " white barred with black." This Lapwing is known alfo in the northern parts of the American continent, appearing in the fpring, and retiring in September: it there lives on berries, infects, and worms.

The ARMED LAPWING of SENEGAL.

THIRD SPECIES.

Parra Senegalla. Linn. and Gmel.
Vanellus Senegalensis Armatus. Briff.
Tringa Senegalla. Lath. Ind.
The Senegal Sandpiper. Lath. Syn.

THIS Senegal Lapwing is as large as the European; but its feet are very tall, and the naked part of its leg meafures twenty lines, and both that part and the feet are greenifh; the bill is fixteen lines long, and bears near the front a narrow membrane, very thin and yellow, hanging down tapered to a point on each fide; the fore part of the body is of the fame colour, but deeper; the great quills of the wing black; thofe next the body of a dirty white; the tail is white in its firft half, then black, and at laft terminating in white. This bird is armed at the fold of the wing with a little horny fpur, two lines in length, and ending in a fhort point.

We may recognize this fpecies in a paffage of Adanfon's Voyage to Senegal, from a habit which belongs, as we have remarked, to the

family

family of the lapwings; that when a perfon appears in their haunts, they flutter about him, and follow his fteps with importunate clamours. Thefe armed lapwings are termed by the French fettlers, *criers (criards)* and by the negroes, *uet-uet.* " As foon as they perceive a man," fays Adanfon, " they fcream with all their force, " and flutter round him, as if to give intima- " tion to the other birds, which, when they " hear the vociferation, make their efcape by " flight: they fpoil, therefore, the fowler's " fport." Our lapwings are peaceful, and never quarrel with other birds; but nature, in beftowing on thofe of Senegal a fpur in the wing, feems to have accoutred them for battle; and they are accordingly faid to employ it as an offenfive weapon.

The ARMED LAPWING of the INDIES.

FOURTH SPECIES.

Parra Goenfis. Gmel.
Tringa Goenfis. Lath. Ind.
The Goa Sandpiper. Lath. Syn.

THIS fpecies was fent us from Goa, and is not yet known to the naturalifts: it is as large as the European lapwing, but taller and more flender; it has a little fpur in the fold of each wing, and its plumage confifts of the ufual colours: the great quills of the wing are black; the tail partly white, partly black, and rufous at the extremity; the fhoulders are covered with a purple tinge; the under fide of the body is white; the throat and the fore-part of the neck are black; the top of the head, and the upper furface of the neck, are alfo black, with a white line on the fides of the neck; the back is brown; the eye is environed by a portion of that exuberant fkin which appears more or lefs in all the armed lapwings and plovers, as if thefe two excrefcences, the fpur and the membranous cafque, had fome common, though concealed, caufe.

The ARMED LAPWING of LOUISIANA.

FIFTH SPECIES.

Parra Ludoviciana. Gmel.
Vanellus Ludovicianus Armatus, Briff.
Tringa Ludoviciana. Lath. Ind.
The Armed Sandpiper. Penn.
The Louifiane Sandpiper. Lath.

THIS is a little fmaller than the preceding, but its legs and feet are proportionally as long, and its fpur is ftronger, and four lines in length: its head is wrapped on each fide with a double yellow band, placed laterally, and which, encircling the eye, is fafhioned behind into a fmall furrow, and ftretches before, on the root of the bill, in two long fhreds: the top of the head is black; the great quills of the wings, too, are black; the tail the fame, with a white point: the reft of the plumage is of a gray ground and tinged with rufty brown, or reddifh on the back, with light reddifh or flefh-colour on the throat and the fore-fide of the neck; the bill and feet are of a greenifh yellow.—We reckon the eighth fpecies of Briffon, denominated *the*

armed

armed lapwing of St. Domingo *, as a variety of the prefent: the proportions are nearly the fame, and the differences feem to refult from age or fex.

---

* Parra Dominica.   *Gmel.*

Thus defcribed by Briffon: " It is dilute fulvous, below inclining
" to rofe-colour; its wing-quills dilute fulvous, the lateral ones in-
" teriorly verging on rofe-colour; a yellow membrane on either fide
" between the bill and the eye, drawn above the eye, and hanging
" downwards, its wings armed."

The ARMED LAPWING of CAYENNE.

SIXTH SPECIES.

Parra Cayanensis. Gmel.
Tringa Cayanensis. Lath. Ind.
The Cayenne Sandpiper. Lath. Syn.

THIS is at least as large as the common lapwing, and is taller: it is also armed with a spur on the shoulder. In its colours it resembles entirely the ordinary species; its shoulder is covered with a mark of bluish gray; a mixture of that colour, with green and purple tints, is spread on the back; the neck is gray, but a broad black space occupies the breast; the front and the throat are black; the tail is partly black, partly white, as in the European lapwing; and, to complete the resemblance, this Cayenne Lapwing has on the back of its head a small tuft of five or six pretty short threads.

It appears, that a species of Armed Lapwing is found also in Chili *; and if the account given by Frezier be not exaggerated, it must

* Parra Chilensis. *Gmel.*

be more ftrongly armed than the reft, fince its
fpurs are an inch long: it is alfo clamorous,
like that of Senegal. " As foon as thefe birds
" fee a man," fays Frezier, they hover round
" him, and fcream, as if to warn the other
" birds, which, at this fignal, fly away on all
" fides."

The LAPWING-PLOVER.

Tringa Squatarola. Linn. and Gmel.
Vanellus Griseus. Briss.
Pluvialis Cinerea. Aldrov. Johnst. Ray and Charleton.
Pardalus. Gesner.
Pluvialis Cinerea Flavescens. Sibb.
Gavia. Klein.
The Gray Plover. Alb. and Brown.
The Gray Sandpiper *. Penn. and Lath.

THIS bird is by Belon termed the *gray-plo-ver*; and in fact it resembles the plover as much as the lapwing, perhaps more: it has indeed, like the latter, the small hind toe, which is wanting in the plover, a difference which has induced naturalists to separate these birds. But it must be observed that this toe is smaller than that of the lapwing, and hardly apparent; and that also it has scarce any of the colours of the lapwing. It might be regarded as a lapwing, because it has a fourth toe; or as a plover, because it has no tuft, and since its habits and its garb are those of the plovers.

Klein will not even admit that this small difference in the toes is a general character, but justly regards it as an anomaly; and he insists

* In Bornholm it is called *Floyte-Tyten, Dolken.*—T.

that

that the lapwings and plovers have fo many
common charaƈters as to conftitute only one
great family.   Accordingly, fome naturalifts
have termed it a lapwing, others a plover : and,
to compromife the matter, and retain the analogies, we have denominated it the *Lapwing-
Plover*.   Fowlers call it the *fea-plover*, which
is an improper appellation, fince it conforts with
the common plovers ; and Belon takes it to be
the leader of their flocks, becaufe it has a louder
and ftronger voice than the reft.   It is fomewhat larger than the golden plover ; its bill *is*
proportionally longer and ftouter : all its plumage is light afh-gray, and almoft white under
the body, mixed with brownifh fpots on the
upper fide of the body and on the fides ; the
quills of the wing are blackifh ; the tail is fhort,
and does not projeƈt beyond the wing.

Aldrovandus conjeƈtures, with a good deal of
probability, that this bird is mentioned by Ariftotle under the name of *pardalis*.   But we muft
obferve that the philofopher does not feem to
fpeak of it as a bird with which he was himfelf
acquainted ; for the following are his expreffions : " The *pardalis* is faid to be in a great
" meafure a gregarious bird, and never found
" alone ; its plumage is entirely cinereous ; it
" is, in point of fize, next the *molliceps* ; its pi-
" nions and feet are vigorous ; its voice not
" deep, but frequent *."   Add to this, that the

* Hift. Animal, *lib.* ix. 23.

name

name *pardalis* fignifies a mottled plumage. All
the other properties belong equally to the family
of the plovers or of the lapwing.

Willughby affures us, that this bird is feen
frequently in the territories of the Venetian
ftate, where it is called *fquatarola*. Marfigli
reckons it an inhabitant of the banks of the
Danube. Schwenckfeld inferts it in the number of Silefian birds; Rzacynfki, in thofe of
Poland; and Sibbald, in thofe of Scotland.
Hence this fpecies, like all the lapwings, is extremely diffufed. Does Linnæus allude to any
peculiarity of its hiftory, when in one of his
editions he denominates it *tringa augufti
menfis* * ? And does it really appear in Sweden
in the month of Auguft ?—The hind toe of
this bird is fo fmall, and fo little apparent, that
with Briffon we fhall not hefitate to refer to it
the *brown lapwing* of Schwenckfeld, though he
fays exprefsly that it wants the hind toe.

To this fpecies alfo we fhall refer, as being
clofely related, the *variegated lapwing* † of
Briffon. Aldrovandus gives the figure without
any defcription; but its appellation fhews that
he knew the great refemblance between the
two birds: all their proportions are nearly the
fame; the ground of their plumage differs only

* *Syft. Nat.* ed. 10. Gen. 60. Sp. 11.
† Tringa-Squatarola, *var. Gmel.*
  Tringa Varia. *Linn.*

in

in a few tints, it being more spotted in this variegated Lapwing. Both of them, according to Brisson, haunt the sea-shore; but it is certain, from the authorities which we have cited, that these birds occur also at a distance from the coast and in inland countries.

[A] Specific character of the Gray Sandpiper: " Its bill is " black, its feet greenish; its body gray, and white below." In England these birds are seen during winter in small and unfrequent flocks. They are observed also in America, flying over the meadows in the back parts of Carolina. They are very common in Siberia. Their flesh is esteemed very delicate.

The P L O V E R S.

LES PLUVIERS. *Buff.*

THE focial inftinct is not beftowed on all the fpecies of birds; but, in thofe which it cements, the union is firmer and more unfhaken than in other animals; not only their flocks are more numerous and more conftantly embodied: the whole community feems to have but one will; and the fame appetites, projects, and pleafures actuate each individual. Birds are more prolific than the quadrupeds, they live in greater plenty, and their motions are performed with greater eafe and celerity. The compactnefs of their fquadrons, and the power of their voice, enable them to transfufe their fentiments and intentions, and to act by mutual concert. And the fagacity exercifed in interpreting their fignals begets among them affection, truft, and the gentle habits of peace and concord. The focieties of quadrupeds, whether formed voluntarily in the wilds of nature, or contracted and upheld by the influence of man, cannot be compared with the congregations of the birds. Pigeons grow fond of their common dwelling; and

and their attachment is the ftronger the more numerous their flock: quails affemble, and concert their migration: the gallinaceous tribes poffefs, even in the favage ftate, thofe focial habits which domeftication only nourifhes and unfolds: laftly, all the birds which fcatter in the woods or difperfe in the fields, gather together in the autumn; and after chearing the bright days of that late feafon with gay fport, they depart embodied, in queft of milder winters and happier climates. All thefe combinations and movements of the feathered race are conducted independently of the guidance or controul of man, though performed under his eye. But his interference in the affociations of quadrupeds difunites and difperfes them. The marmot, formed by nature for fociety, now lives folitary and exiled on the fummits of mountains; the beaver, difpofed ftill more to friendfhip, and almoft civilized, has been driven into the deepeft wilds. Man has deftroyed or prevented all union among animals: that of the horfe has been extinguifhed, and the whole fpecies fubjected to the rein*: even the elephant has been

reduced

---

* The horfes which have grown wild on the plains of Buenos-Ayres go in large flocks, run together, feed together, and give all the marks of mutual attachment and intelligence, and of delighting in fociety. The fame is the cafe with the wild dogs in Canada and other parts of North America. We can fcarce doubt but that the other domeftic fpecies, that of the camel, fo long reduced to fubjection; that of the ox and of the fheep, which man

has

reduced to conftraint, notwithftanding his vaft
ftrength, and his conftant fterility in the domeftic ftate. The birds alone have efcaped the do-
mination of the tyrant; and their fociety is as
free as the element which they inhabit. His
attacks can deftroy only the life of the individual; the fpecies may fuffer a diminution of its
numbers, but its inftincts, habits, and œconomy
remain untouched. There are many fpecies
even which are known to us only from their
focial propenfity, and are never feen but at the
time of their general mufter, when vaft multitudes are affembled. Such in general are the
companies of many water-fowls, and in particular that of the Plovers.

They appear in numerous bodies in the provinces of France during the autumnal rains,
and from this circumftance they derive their
name *. They frequent, like the lapwings, the
wet bottoms and flimy grounds, where they
fearch for worms and infects: they go into the
water in the morning to wafh their bill and feet,
which are clotted with mud by their employ-
ment; a habit which is common alfo to the
woodcocks, the lapwings, the curlews, and many

has difunited by degrading them with fervitude; were alfo naturally focial, and difplayed in the wild ftate, ennobled by freedom,
thofe tender tokens of regard and affection with which we behold
them mutually foothe their flavery.

* From *pluvia*, rain. Gefner fuppofes it to come from *pulvis*,
duft; which is much lefs probable, there being many other birds
befides Plovers that welter in duft.

other

other birds which feed on worms. They ftrike
the ground with their feet to elicit thefe, and
often they extract them from their retreat *.
Though the Plovers are ufually very fat, their
inteftines are found to be fo empty, that it has
been fuppofed that they could live on air; but
it is probable that the foft fubftance of the
worms turns wholly into nourifhment, and
leaves little excrement. They feem however
capable of fupporting a long abftinence:
Schwenckfeld fays, that he kept one fourteen
days, which during the whole time only drank
fome water, and fwallowed a few grains of
fand.

Seldom do the Plovers remain more than
twenty-four hours in the fame place: as they
are very numerous, they quickly confume the
provifions which it affords, and are then obliged
to remove to another pafture. The firft fnows
compel them to leave our climates, and feek
milder regions: however, a confiderable number of them remain in our maritime provinces †
till the hard frofts. They return in fpring ‡,
and always in flocks; a fingle Plover is never

* Note communicated by M. Baillon, of Montreuil-fur-mer.

† In Picardy, according to M. Baillon, many of thefe birds
continue in the neighbourhood of Montreuil-fur-mer, till the intenfe frofts fet in.

‡ The Chevalier Defmazy informs us, that they are feen to pafs
Malta regularly twice a year, in fpring and in autumn, with a
multitude of other birds which crofs the Mediterranean, and make
that ifland their place of ftation and repofe.

*

feen, fays Longolius. And, according to Belon, their fmalleft companies amount at leaft to
fifty. When on the ground they never reft, but
are inceffantly engaged in the fearch of food;
they are almoft perpetually in motion : feveral
keep watch while the reft of the flock are feeding, and on the leaft fymptom of danger they
utter a fhrill fcream, which is the fignal of
flight. On wing, they follow the wind, and
maintain a pretty fingular arrangement; and
thus advancing in front, they form in the air
tranfverfe zones, very narrow and exceedingly
long: fometimes there are feveral of thefe zones
parallel, of fmall depth, but wide extended in
crofs lines.

When on the ground thefe birds run much,
and very fwiftly; they continue in a flock the
whole day, and only feparate to pafs the night :
they difperfe in the evening to a certain haunt,
where each repofes apart; but at day-break,
the one firft awake or the moft watchful, which
fowlers term the *caller*, though perhaps it is the
fentinel, founds the cry *hui, hieu, huit*, and in an
inftant they obey the fummons and collect together. This is the time chofen for catching
them : a clap-net is ftretched before dawn fac-
ing the place where they fleep; a number of
fowlers encircle it, and as foon as the call is
heard, they throw themfelves flat on the ground
till the birds gather; then they rife up, fhout,
and throw their fticks into the air; fo that the
                                                    Plovers

Plovers are frightened, and hurrying away with a low flight they ftrike againft the net, which drops upon them, and often the whole flock is taken. This plan is always attended with great fuccefs; but a fingle bird-catcher can in a fimpler way enfnare confiderable numbers: he conceals himfelf behind his net, and attracts the birds by means of a call of bark. They are caught in abundance in the plains of Beauce and of Champagne. Though very common in Italy, they are efteemed excellent game: Belon fays, that in his time a Plover was fold often as dear as a hare; he adds, that they preferred the young ones, which he calls *guillemots*.

The chafe of the Plovers, and their mode of living in that feafon, are almoft the whole we know of their natural hiftory. Tranfient guefts rather than inhabitants of our fields, they difappear on the fnow's falling; repafs without halting in the fpring, and leave us when the other birds arrive. It would feem, that the gentle warmth of that feafon, which awakens the dormant faculties of the other birds, makes a contrary impreffion on the Plovers: they proceed to the more northern countries to breed, and rear their young, for during the whole fummer we never fee them. Then they inhabit Lapland, and other parts of the north of Europe *, and

* See Collection Academique, *partie etrangere*, tome xi. *Academie de Stockholm*, p. 60.

probably

probably thofe of Afia. Their progrefs is the fame in America; fince they are common to both continents, and are obferved in the fpring at Hudfon's Bay advancing farther north *. After arriving in flocks in thofe arctic tracts, they feparate into pairs; and the more intimate union of love breaks, or rather fufpends for a time, the general fociety of friendfhip. Hence Klein, an inhabitant of Dantzick, remarks, that the Plovers live folitary in low grounds and meadows.

The fpecies which in our climates appears as numerous at leaft as that of the lapwing, is lefs diffufed. According to Aldrovandus, fewer Plovers are caught in Italy than lapwings, and they are not found in Switzerland and other countries, where the lapwings are frequent. But perhaps the Plover, advancing farther to the north than the lapwing, gains as much territory as it relinquifhes in the fouth. It feems alfo to have occupied a fpacious tract in the new world, which has afforded an ample range to many fpecies of birds, becaufe there the temperature is more uniform throughout, and the climates more obfcurely difcriminated.

The Golden Plover may be regarded as the reprefentative of the whole family of Plovers; and what we have faid of their habits and œconomy refer to it: but fpecies are included, which we proceed to enumerate and defcribe.

* Hift. Gen. des Voy. *tom.* xv. *p.* 267.

The GOLDEN PLOVER.

LE PLUVIER DORE'. *Buff.*

FIRST SPECIES.

Charadrius-Pluvialis. Linn. and Gmel.
Gavia Viridis. Klein.
Pluvialis Viridis. Ray. Will. and Sibbald.
Pluvialis Aurea. Briff.
Pivier. Aldrov.
Pluvialis Flavefcens. Johnft.
Pluvialis Flavo-Virefcens. Charleton.
* *The Golden, or Green Plover.* Penn. Lath. &c.

THE Golden Plover is as large as the turtle;
its length from the bill to the tail, and alfo
from the bill to the nails, is about ten inches:
all the upper fide of the body is dafhed with yellow ftreaks, intermixed with light-gray, on a
blackifh brown ground; and thefe yellow ftreaks
are confpicuous in the dark field, and give the
plumage a golden luftre. The fame colours,
only more dilute, are intermingled on the throat
and breaft; the belly is white, the bill is black,

---

* In German *Pluvier, Pulrofz, See-taube* (fea-pigeon), *Gruner-
kiwit* (green pewit): in Italian *Pivrero, Piviero Verde:* in Polifh
Ptak-deffezowy: in Swedifh *Aokerhoens:* in Norwegian *Akerloe:*
in Lapponic *Hutti:* in Catalonia it is called *Dorada;* and in Silefia *Brach-vogel.*

as

THE GOLDEN PLOVER.

as in all the plovers, fhort, rounded, and fwelled
at the tip; the feet are blackifh, and the outer
toe is connected as far as the firft joint, by a
fmall membrane, to the mid-toe; the feet have
only three toes, and there is no veftige of a hind-
toe or heel; which property, joined to the inflation of its bill, is regarded by ornithologifts
as the difcriminating character of the plovers.
In all of them, a part above the knee is bare;
the neck fhort; the eyes large; the head rather
too bulky in proportion to the body: qualities
which belong alfo to all the *fcolopacious* birds *,
which fome naturalifts have ranged together
under the denomination of *pardales* †; though
there are many fpecies, particularly among the
plovers, whofe plumage is not mottled like a
panther or tiger.

There is little difference between the plumage of the male and that of the female ‡:
however, the varieties, whether individual or
accidental, are very frequent; fo that in the
fame feafon, out of five-and-twenty or thirty
Golden Plovers, we fhall hardly find two exactly alike. They have more or lefs of yellow,
and fometimes fo little of it as to appear quite
gray §. A few have black fpots on the breaft,
&c.

* As the woodcocks, the fnipes, the godwits, &c.
† Klein, Schwenckfeld.
‡ Aldrovandus, Belon.
§ M. Baillon, who has obferved thefe birds in Picardy, affures
us that their early plumage is gray; that at their firft moult, in
August

&c. These birds, according to M. Baillon, arrive on the coasts of Picardy about the end of September or the beginning of October; but in our more southern provinces they do not appear until November, or even later, and they retire in February and March *. In summer, they are seen in the north of Sweden, in Dalecarlia, and in the isle of Oëland †, in Norway, Iceland, and Lapland ‡. From these arctic regions they appear to have migrated into the new world, where they seem to be more widely diffused than in the old; for a Golden Plover, differing only in some shades from the European, is found in Jamaica §, in Martinico, in St. Domingo ||, and in Cayenne. In the southern parts of Ame-

---

August and September, they get some feathers, which have a yellow cast, or which are spotted with that colour; but that it is not till after some years that they acquire their fine golden tint. He adds, that the females are hatched entirely gray, and long retain that colour; that it is only when they grow old that they assume a little yellow; and that it is very rare to see them have their plumage so uniform and beautiful as that of the males. Thus we need not wonder at the variety of colours remarked in this species of birds, since they result from the difference of age and of sex. *Note communicated by M. Baillon.*

* M. Lottinger has observed the same of their passage in Lorraine.

† *Fauna Suecica.*

‡ Brunnich.

§ Sloane.

|| Charadrius-Pluvialis, *var.* z. *Gmel.*

Thus described by Brisson: " Above, blackish, variegated with " yellowish spots; below white; the lower part of its neck and its " breast dilute gray; the edges of its quills yellowish; its tail-quills " brown, spotted at the edges with yellowish white."

rica

rica thefe birds inhabit the favannas, and vifit the patches of fugar-cane which have been fet on fire. Their flocks are numerous, and can hardly be approached: they are migratory, and are feen in Cayenne only during the rainy feafon.

M. Briffon eftablifhes a fecond fpecies, which he denominates *the leffer Golden Plover* *, and as his authority, cites Gefner, who never faw the Plover himfelf. Schwenckfeld and Rzacynfki alfo mention this fmall fpecies, and probably ftill from Gefner; for the former, though he applies to it the epithet ' little,' fays, at the fame time, that it is equal in bulk to the turtle; and the latter adds no particulars that imply that he obferved it diftinctly himfelf. We fhall therefore confider this little Golden Plover as only an individual variety.

---

* Charadrius-Pluvialis. *var.* 1. *Gmel.*
Thus defcribed by Briffon : " Above blackifh, variegated with " yellowifh fpots; below white; the tail-quills blackifh, fpotted at " the edges with yellowifh white."

[A] Specific character of the Golden Plover, *Charadrius-Pluvialis* : " Its body is fpotted with black and green; below " whitifh; its feet cinereous." It lays four eggs, about two inches long, fharper than thofe of the lapwing, of a pale olive, variegated with blackifh fpots. It is often found in the winter fea-fon on our moors and heaths in fmall flocks.

The ALWARGRIM PLOVER.

LE PLUVIER DORE A GORGE NOIRE.

Buff. *

SECOND SPECIES.

Charadrius Apricarius. Linn. and Gmel.
Pluvialis Aurea Freti-Hudfonis. Briff.
The Hawk's-eye spotted Plover. Edw. and Bancr.

THIS fpecies is often found with the preceding in the northern countries, where
they live and propagate, but without intermix-
ture. Edwards received this bird from Hudfon's Bay, and Linnæus met with it in Sweden,
in Smoland, and in the wafte plains of Oëland †:
it is the *Pluvialis minor Nigro-flavus* ‡ of Rud-
beck. Its front is white, and a fmall white
fillet, paffing over the eyes and the fides of the
neck, defcends before, and encircles a black
mark which covers the throat : the reft of the
under furface of the body is black : all the
mantle is dufky brown and blackifh, and fpeckled pleafantly with a vivid yellow, which is dif-
tributed by indented fpots on the margin of

* *i. e.* Black-throated Golden Plover.
† In Smoland it is called *Myrpitta,* and in Oëland *Alwargrim.*
‡ *i. e.* The black-yellow leffer plover.

each

each feather. This Plover is as large as the preceding. We know not why the Englifh fettlers at Hudfon's Bay give it the epithet of *Hawk's-eye* ; whether by antiphrafis they allude to its weak eyes, or really fignify that its fight is fuperior to that of other birds of its kind.

[A] Specific charaƈter of the Alwargrim Plover, *Charadrius Apricarius* : " Its throat and belly are black ; its body dotted with " brown, white, and yellow ; its feet cinereous." It appears in Greenland in the fpring, and lives on worms and heath-berries. In North America it breeds, and fpends the fu nmer months in the northern ftates. Its brilliancy has procured it the name of *Hawk's- eye* in Hudfon's Bay. Its flefh is reckoned delicious.

The D O T T E R E L.

LE GUIGNARD. *Buff.*

THIRD SPECIES.

Charadrius-Morinellus. Linn. and Gmel.
Pluvialis Minor, five Morinellus. Briff.
Morinellus. Sibb. Charl. Will. &c.
Morinellus Anglorum. Gefner.

THIS bird is by fome called the *little plover*. It is indeed fmaller than the golden plover, not exceeding eight inches and a half in length: the ground of its upper furface is brown-gray with a green glofs; every feather of the back, and alfo the middle ones of the bill, are border- ed with a rufous ftreak; the upper part of the head is blackifh brown; the fides and the face are fpotted with gray and white; fore part of the neck and the breaft are undated gray, rounded into a mark, under which, and near a black ftreak, there is a white zone, which is the diftinguifh- ing character of the male: the ftomach is ru- fous; the belly black; and the abdomen white.

The Dotterel is well known for the excel- lence of its flefh, which is ftill more delicate and juicy than that of the golden plover. The fpecies

cies feems to be more difperfed in the north
than in our climates; and, beginning with
England, it extends to Sweden and Lapland *.
This bird has two annual flittings, in April and
in Auguft; in which it removes from the marfhes
to the mountains, attracted by the black beetles,
which are its chief fubfiftence, together with
worms and fmall land cockles, which are found
in its ftomach †. Willughby defcribes the method of catching them practifed in the county of
Norfolk, where they are numerous: five or fix
fportfmen fet out together, and when they difcover the birds, they ftretch a net at fome diftance beyond them; then they advance foftly,
throwing ftones or bits of wood, and the indolent birds, thus roufed from their fleep, ftretch
out one wing or one foot, and can fcarce ftir:
the fowlers believe that they mimic whatever
they fee, and therefore endeavour to amufe them
by extending their arm or their leg, and by this
manœuvre, apparently idle ‡, to draw off their
attention: but the Dotterels approach flowly

* In the fixth edition of the *Syftema Naturæ*, it is denominated
Charadrius Lapponicus.

† Letter of Dr. Lifter to Mr. Ray. *Philofophical Tranfactions*,
N° 175. Art. 3.

‡ An author, in Gefner, goes fo far as to fay, that this bird, attentive to the motions of the fowler, and delighted as it were, imitates all his geftures, and forgets its own prefervation, infomuch as
to fuffer him to approach and cover it with the net which he holds
in his hand. See Aldrovandus, tom. iii. p. 540.

and

and with a fluggish pace to the net, which drops and covers the stupid troop.

This character of fluggishnefs and ftupidity has given occafion to the English name *Dotterel*, and alfo to the Latin appellation *Morinellus* *. Klein fays, that its head is rounder than that of any of the plovers, which he reckons a mark of their dullnefs, from the analogy to the round heads of the breed called the *foolish pigeons*. Willughby thought he could perceive, that the females were rather larger than the males, without any other exterior difference.

With regard to the fecond fpecies, which Briffon reckons, of the Dotterel, under the name of the *English Dotterel*, though both birds inhabit England; we fhall confider it as merely a variety. Albin reprefents it too fmall in his figure, fince in his defcription he affigns greater weight and meafures than to the common Dotterel: indeed the chief difference confifts in this, that it wants the crofs bar below the breaft, and that the whole of that part, with the ftomach and the fore fide of the neck, are lightgray wafhed with yellowifh. It appears to me therefore unneceffary to multiply fpecies on fuch flight foundations.

* *Dotterel* derived from the verb *to dote*. *Morinellus* formed from *Morio*, a fool or jefter.

[A] Specific character of the Dotterel, *Charadrius-Morinellus*: " Its breaft ferruginous ; a white linear bar on its eye-brows and " breaft ; its feet black." " Thefe birds," fays Mr. Pennant, " are " found

" found in Cambridgeſhire, Lincolnſhire, and Derbyſhire. On
" Lincoln-heath, and on the moors of Derbyſhire, they are migra-
" tory, appearing there in ſmall flocks of eight or ten only in the
" latter end of April, and ſtay there all May and part of June, dur-
" ing which time they are very fat, and much eſteemed for their
" delicate flavour. In the months of April and September they
" are taken in Wiltſhire and Berkſhire downs.—At preſent, ſportſ-
" men watch the arrival of the Dotterels, and ſhoot them ; the
" other method (that deſcribed in the text) having been long dif-
" uſed."

The R I N G   P L O V E R.

LE PLUVIER 'A COLLIER. *Buff.* *

F O U R T H   S P E C I E S.

Charadrius, seu Hiaticula †.   Ald. Johnst. Sibb.

WE shall divide this species into two branches; the first is as large as a red-wing ‡,
the second nearly equal to a lark §.   And the
latter must be understood to represent the Ring

* *i. e.* the Collared Plover.

† In Polish *Zoltaczek:* in Swedish *Strand-pipare, Grylle, Trulls:*
in Lapponic *Pago:* at Bornholm *Præste-Krave, Sand-Vrifter:* in
Brasilian *Matuitui.*

‡ *Charadrius Alexandrinus.*   Linn. and Gmel.
Charadrius Ægyptius.   Linn.
Pluvialis Torquata.   Briss.
Gavia Littoralis.   Klein.
The Alexandrine Plover.   Lath.

Specific character: " It is brown; its front, the collar on its
" back, and its belly, white; its lateral tail-quills on both sides
" bright white; its feet black."

§ *Charadrius-Hiaticula.*   Linn. and Gmel.
Pluvialis Torquata Minor.   Briss.
The Sea Lark.   Alb. Will. and Sloane.
The Ringed Plover.   Penn. and Lath.

Specific character: " Its breast brown, its front blackish with a
" white ring, its top brown, its feet yellow."

Plover,

THE RINGED PLOVER.

Plover, as it is more diffused and better known than the former, which is perhaps only a variety.

Their head is round, their bill very short, and thick feathered at the root; the first half of the bill is white or yellow, and the tip is black; the front is white; the crown of the head has a black band, and a gray cap covers it; this cap is edged with a black fillet, which rifes on the bill and passes under the eyes; the collar is white; the mantle is brown gray; the quills of the wing are black; the under side of the body is a fine white, and also the front and the collar.

Such is in general the plumage of the Ring Plover; but to describe all the diversities in the diftribution and intensity of the colours were endless. Yet notwithftanding thefe local or individual differences, the bird is the fame in almost all climates. It is brought from the Cape of Good Hope, from the Philippines *, from Louisiana, and from Cayenne †. Captain Cook found it in the ftraits of Magellan ‡, and Ellis, at Hudson's Bay §. It is the fame with what Marcgrave calls the *matuitui* of Brazil. Willughby makes that remark, and expresses his

---

* Sonnerat. *Voyage à la Nouvelle Guinée*, p. 83.

† At Cayenne it is called *collier*; and the Spaniards of St. Domingo, feeing it robed in black and white like their monks, termed it *frailecitos*; the Indians give it the name *thegle, thegle,* from its cry. *Feuillée.*

‡ At Famine Bay.

§ Near Nelfon River.

furprize

furprize that there fhould be birds common to
South America and to Europe: a fact extraordinary and inexplicable, except on the prin-
ciple which we eftablifh in treating of the water fowls; that the element which they inhabit
is in all latitudes nearly of an equal temperature,
and every where yields the fame fubfiftence.
We fhall therefore regard the Ring Plover as
one of thofe fpecies which are fpread over the
whole globe, and derive the varieties which occur in the plumage from the influence of climate *.

The Ring Plovers inhabit the verge of waters; they are obferved on the fea-fhore following the tide. They run very nimbly in the
ftrands, at times taking fhort flights, and always fcreaming. In England, their nefts are
found on the rocks by the coaft: there they are
very common, as in moft of the northern countries; in Pruffia †, Sweden ‡, and ftill more
in Lapland during fummer. A few of thefe
birds are found alfo on the rivers, and in fome
of the provinces of France: they are called
gravieres (channel-birds) and in other places

* We reckon the *Greateft Snipe* of Sloane and Ray one of the
varieties. It is the *Pluvialis Jamaicenfis Torquata* of Briffon, and
the *Charadrius Jamaicenfis* of Gmelin, who thus characterifes it:
" Above brown, below white; its breaft black and white; its tail
" whitifh, variegated with rufous and blackifh; its collar and its
" feet black."
† Rzaczynfki.
‡ Linnæus.

6

criards

criards (fcreamers) which they well merit for their troublefome and continual cries during the education of their young, which lafts fo long as a month or fix weeks. Fowlers affure us, that they make no nefts, but drop their eggs on the gravel, and that thefe are greenifh fpotted with brown. The parents lurk in holes under the projecting brinks * ; and hence ornithologifts have inferred it to be the fame with the *charadrios* of Ariftotle, which, as the word imports, was an inhabitant of *channels*, or *gullies* † ; and whofe plumage, the philofopher adds, is as difagreeable as its voice. Ariftotle alfo fays, that it comes abroad at night, and lies concealed during the day ‡ ; this remark, though not precifely applicable to the Ring Plover, has perhaps fome relation to its habits, fince it is heard very late in the evening. The charadrios was one of thofe birds to which ancient medicine or rather fuperftition afcribed occult virtues, and it was fuppofed to cure the jaundice: the patient needed only to look at the bird §, which at the fame time turned afide its eyes as if affected by

---

* Klein.

† Ariftophanes gives the charadrios the office of conveying water into the city of the birds.

‡ Hift. Animal, *lib.* ix. 11.

§ The vender of this excellent remedy was careful to conceal his bird, felling only the fight of it: this gave occafion to a proverb among the Greeks, applied to thofe who kept any thing precious or ufeful concealed, *imitating a charadrios.* Gefner.

the

the diforder *.   What imaginary remedies has
human weaknefs fought for its real ills !

* Heliodorus, *Æthiopic. lib.* iii.

[A] The Ring Plovers are common on the Britifh coafts in
fummer, but difappear on the approach of winter.

The NOISY PLOVER.

LE KILDIR. *Buff.*

FIFTH SPECIES.

Charadrius Vociferus. Linn. and Gmel.
Pluvialis Virginiana Torquata. Briff.
The Chattering Plover, or Kill-Deer. Catefby.

THE name *Kill-Deer*, which this bird has in
Virginia, is expreffive of its cry. It is very
common in that province and in Carolina, and
is detefted by the fowlers, becaufe its clamours
fcare away every other fort of game. There
is a good figure of this bird in Catefby's work:
it is as large, he fays, as the fnipe; its legs tall;
all its upper furface is brown-gray; and the top
of its head is hooded with the fame colour; its
front, its throat, the under fide of its body, and
the compafs of its neck, are white; the lower
part of the neck is encircled by a black collar,
below which is a white half collar: there is alfo
a black bar on the breaft, which ftretches from
the one wing to the other; the tail is pretty
long, and black at the extremity; the reft of it,
and the fuperior coverts, are of a rufous colour;
the feet are yellowifh; the bill is black; the eye

is

is large, and environed with a red circle. Thefe
birds remain the whole year in Virginia and Carolina ; they are found too in Louifiana *. No
difference can be perceived in the plumage between the male and the female.

A fpecies akin to this, perhaps the fame, is
the collared plover of St. Domingo †, which
requires no other defcription ; the only difference lies in the colours of the tail, and the
deeper tint of the wings.

* Dr. Mauduit has received it from that country, and preferves
it in his cabinet.

    † *Charadrius Vociferus, var.* Gmel.
      Charadrius Torquatus.   Linn.
      Pluvialis Dominicenfis Torquata.   Briff.

Briffon fays, " that the two middle quills of the tail are gray-
" brown, tawny at the tip ; the two next on either fide gray-brown,
" black near the tip, and the tip itfelf white ; the outmoft white at
" its origin, ftriped tranfverfely with blackifh."

[A] Specific character of the Noify Plover, *Charadrius Vocife-*
rus : " It has black bars on its breaft, its neck, its front, and its
" cheeks ; its tail is yellow, with a black bar ; its feet bright yel-
" low."

The CRESTED PLOVER.

LE PLUVIER HUPPÉ. *Buff.*

SIXTH SPECIES.

Charadrius Spinosus, var. Linn.
Pluvialis Persica Cristata. Briss.
The Black-breasted Indian Plover. Edw.

THIS Plover, which is found in Persia, is nearly as large as the golden plover, but somewhat taller; the feathers on the crown of the head are black, glistening with green; they are collected into a tuft, which reclines, falls back, and is about an inch long; the cheeks, the back of the head, and the sides of the neck, are marked with white; all the upper surface is deep chesnut; a black streak descends from the throat upon the breast, which, as well as the stomach, is black, with a fine violet glofs; the lower belly is white; the tail is white at its origin, and black at its extremity; the quills of the wing, too, are black, and the great coverts are marked with white.

This Plover is armed with a spur on the wing. The female is distinguished from the male; all its throat being white, and its black not shaded by any admixture.

The SPUR-WINGED PLOVER.

Le Pluvier a Aigrette. *Buff*.

SEVENTH SPECIES.

Charadrius Spinosus. Linn. an⁻ Gmel.
Pluvialis Senegalensis Armata. Briss.

T HE feathers on the back of the head extend into threads, as in the lapwing, and form a tuft more than an inch long: it is of the bulk of the golden plover, but taller, measuring a foot from the bill to the nails, and only eleven inches from the bill to the end of the tail; the top of the head, the tuft, the throat, and the mark on the stomach, are black, and also the great feathers of the wing, and the tips of those of the tail; the upper surface is brown gray; the sides of the neck, the belly, and the great coverts of the wing, are white tinged with fulvous: the spur on the fold of the wing is black, strong, and six lines long. This species is found in Senegal, and occurs too in the hot parts of Asia, for we received one from Aleppo.

[A] Specific character of the Spur-winged Plover, *Charadriu⁻ Spinosus:* " Its wing-quills, its breast, and its feet, are deep black; " the back of its head crested; its tail-quills half white; its bas- " tard wings armed with spurs."

The HOODED PLOVER.

Le Pluvìer Coiffe'. *Buff.*

EIGHTH SPECIES.

Charadrius Pileatus. Gmel.

A PORTION of yellow membrane paffes on the front, and by its extenfion encircles the eye; a black hood, ftretched behind into two or three fhreds, covers the top of the head; the nape is white, and a broad black chin-piece, rifing under the eye, envelopes the throat, and encircles the top of the neck: all the fore fide of the body is white; the upper fide of the body is rufty gray; the quills of the wing, and the end of the tail, are black; the feet are red, and there is a fpot of the fame colour near the point of the bill. This Plover is found in Senegal, like the preceding, but is a fourth fmaller, and has no fpur in the wing. The fpecies is new.

The CROWNED PLOVER,

NINTH SPECIES.

Charadrius Coronatus. Gmel.
The Wreathed Plover. Lath.

THIS is one of the largeft of the Plovers; it is a foot long, and its legs are taller than the golden plover; they are rufty-coloured; the head is hooded with black, in which there is a white band, which encompaffes the whole of the head, and forms a fort of crown; the fore fide of the neck is gray, and the gray colour of the breaft is intermingled with black in coarfe waves; the belly is white; on the firft half of the tail, and at its extremity, a black bar croffes the white; the quills of the wing are black, and the great coverts white; all the upper furface is brown, gloffed with greenifh and purple. It is found at the Cape of Good Hope.

The WATTLED PLOVER.

LE PLUVIER A` LAMBEAUX. *Buff.*

TENTH SPECIES,

Charadrius Bilobus. Gmel.

THE character from which we have derived the name of this bird, is a yellow membrane adhering to the corners of the bill, and hanging from the two fides in pointed wattles. It is found in Malabar : it is of the bulk of the ordinary plover, but its legs are taller, and of a yellowifh caft; behind the eyes there is a white ftreak, which edges the black hood of the head; the wing is black, and fpotted with white on the great coverts; there is alfo black bordered with white on the tip of the tail; the upper furface, and the neck, are fulvous gray, and the under fide of the body is white : this is the common, and, we might fay, the uniform plumage of moft of the Plovers.

The ARMED PLOVER of CAYENNE.

Charadrius Cayanus. Lath.

THIS is a Ring Plover, like the common one, but much taller: its bill, too, is longer, and its head not fo round; a broad black band covers the front, inclofes the eyes, and joins into the black which ftains the back of the neck, the top of the back, and gathers into a round mark on the breaft: the throat is white, and alfo the fore fide of the neck, and the under fide of the body; a gray fpace, with a white edging, forms a hood behind the head; the firft half of the tail is white, and the reft is black; the quills of the wings and the fhoulders are black alfo; the reft of the upper furface is gray mixed with white; the fpurs are pretty long, and inferted in the fold of the wings.

It appears to us, that the *amacozque* of Fernandez, *a noify bird, the plumage mixed with white and black, and a double collar, which is feen the whole year on the lake of Mexico, where it lives on aquatic worms,* is a Plover: we could have decided the queftion, had Fernandez given the character of the feet.

The PLUVIAN.

Charadrius Melanocephalus. Gmel.
The Black-headed Plover. Lath.

I T is fcarce equal to the little ringed plover,
but its neck is longer and its bill ftronger:
the upper fide of the head, of the neck, and of
the back, is black; and there is a ftreak of the
fame colour on the eyes, and fome black waves
on the breaft: the great quills of the wing are
mixed with black and white: the other parts of
the wing, the middle quills and coverts, are of an
handfome gray; the fore fide of the neck is
rufty white, and the belly white; it has three
toes, like the plover, but the bill is larger and
thicker, and the inflation is more perceptible.
Thefe differences feem to conftitute a fhade in
the genus, and I have therefore given it a diftinct name.

The GREAT PLOVER*.

Charadrius-Oedicnemus. Linn. and Gmel.
Pluvialis Major. Briff.
Otis-Oedicnemus. Lath.
Oedicnemus. Belon, Aldrov. and Johnft.
Charadrius. Gefner.
Fedoa Noftra Tertia. Ray.
The Stone Curlew. Will.
The Norfolk Plover. Penn.
The Thick-knee'd Buftard. Lath.

THERE are few perfons who refide in the
provinces of Picardy, Orleanois, Beauce,
Champagne, and Burgundy, but muft have
heard in the fields, in the evening, during the
months of September, October, and November,
the repeated cries *tũrrlui, tũrrlui,* of thefe
birds: this is their call, which often re-echoes
from hill to hill; and as it refembles the articulated found of the curlew, it has probably
given occafion to the appellation of land curlew *(courlis de terre.)* Belon fays, that at firft
fight it appears fo like the little buftard, that he
gave it the fame name. But it is neither a curlew nor a buftard: it is rather a plover, though

---

* In France it is commonly called *Courlis de Terre*, and on the coafts
of Picardy *St. Germer*. In Italy it is termed *Coruz*, according to
Gefner and Aldrovandus; and at Rome *Carlotte*, according to Willughby: in fome parts of Germany *Triel*, or *Griel*, according to
Gefner.

I

it

THE THICK KNEED BUSTARD.

it has feveral peculiar features and habits, that disjoin it from the other fpecies.

This bird is much larger than the golden plover, and even exceeds the woodcock: its thick legs have a remarkable fwelling below the knee; for which reafon Belon has applied the epithet *œdicnemus* *. Like the plover it has only three toes, which are very fhort: its legs and feet are yellow; its bill yellowifh from the origin to the middle, and from thence blackifh to the extremity; and is of the fame fhape with that of the plover, only thicker: all the plumage is of a light gray and rufty-gray ground, and fpeckled with dafhes of brown and blackifh, the ftrokes very diftinct on the neck and breaft, and more confufed on the back and the wings, which are croffed with a whitifh bar: two ftreaks of rufty white pafs above and below the eye: the ground is rufty colour on the back and neck, and white under the belly, which is not fpeckled.

This bird has great power of wing; it fprings at a diftance, efpecially in the day-time, and flies pretty near the ground: it runs on the fward and in the fields, as fwiftly as a dog; and hence, in fome provinces, it has been termed *the furveyor* (arpenteur.) After running, it ftops fhort, holding its body and head ftill †; and on

* Formed from οιδεω, to fwell, and κνημη, the upper part of the leg or knee.
† Albin.

the

the leaſt noiſe it ſquats on the ground. Flies
beetles, ſmall ſnails, &c. are its chief food, together with ſome other inſects that are found in
fallow grounds, ſuch as crickets, graſshoppers,
&c. *: for it conſtantly inhabits the brows of
banks, and prefers the ſtony, ſandy, and dry
ſpots. In Beauce, Salerne tells us, bad land is
called *curlew-land*. Theſe birds, ſolitary and
tranquil during the day, begin to ſtir on the approach of the evening ; then they ſpread on all
ſides, flying rapidly, and crying on the heights
with all their might: their voice, which is heard
at a great diſtance, reſembles the ſound of a
third flute, dwelling on three or four tones, from
a flat to a ſharp. This is the time that they
approach our dwellings †.

Theſe nocturnal habits ſeem to ſhew, that
this bird ſees better in the night than in the
day ; yet is its ſight very acute in the day time:
the poſition too of its large eyes enables it to
ſee as well before as behind : it deſcries the
ſportſman at a good diſtance, ſo that it can riſe
and eſcape before he gets within ſhot of it. It
is as wild as it is timorous ; fear alone confines
it during the day, and permits it not to come
abroad or utter its cries till night : and ſo invin-
cible is its timidity, that if a perſon enter the

* M. Baillon, who has obſerved this bird on the coaſts of Picardy,
informs us, that it alſo eats little black lizards, which it finds on
the downs, and even ſmall adders,

† Sloane.

room

room where it is kept, it endeavours to conceal itself, or sneak off, and runs against whatever happens to be in its way. It is said, that this bird foresees the changes of the weather, and announces rain: Gesner remarks, that even when confined, it is much disturbed before a storm.

This Great Plover, or Stone Curlew, forms an exception to the numerous species, which having a part of their leg naked, are reckoned inhabitants of shores and boggy grounds; since it keeps ever at a distance from water or marshes, and resides in dry upland situations *.

These are not the only habits in which they differ from the plovers. The seasons of their migrations are not the same; for they retire in November, during the latter rains of autumn, having, previous to their departure, which commences at night †, collected at the call of their leader, in flocks of three or four hundred ; and they return early in the spring, being seen in the end of March in Beauce, Sologne, Berry, and some other provinces of France. The female lays only two, or sometimes three eggs, on the naked ground, between stones ‡, or in a small hole which she forms in the sand of heaths or

---

* We may hence see with what little reason Gesner took it for the *charadrios* of Aristotle, which is decidedly a shore bird. See the article of the *Ring-Plover*.

† Salerne.

‡ Idem.

downs.

downs *. The male purfues her eagerly in the love feafon; he is as conftant as he is ardent, and never deferts her: he affifts in leading the young, and inftructing them to diftinguifh their food. The education is tedious; for though the brood can foon walk after they are hatched, it is a confiderable time before they have ftrength fufficient to fly. Belon found fome that could not ufe their wings in the end of October; which induced him to fuppofe that the eggs are laid at a late feafon. But the Chevalier Defmazy, who obferved thefe birds at Malta †, informs us, that they have regularly two hatches, one in the fpring, and another in the end of Auguft. He alfo affures us, that the incubation lafts thirty days. The young ones are excellent game, and the adults likewife are eaten, though their flefh is blacker and dryer. The chafe of

* During the eight days which I wandered on the dry fands that cover the fea-fhore from the mouth of the Somme to the extremity of the Boulonnois, I met with a neft which appeared to be that of the *faint-germer* (ftone-curlew). To afcertain this, I remained fitting on the fand till evening, having raifed a little hillock in front to conceal myfelf. The birds which inhabit thofe fands being accuftomed to fee the furface perpetually changed by the wind, are not in the leaft difturbed at finding new cavities or elevations. My trouble was recompenfed: in the evening the bird came to her eggs, and I recognized it to be the *faint-germer*, or ftone-curlew: her neft placed expofed on a flat in a fand-plain, confifted only of a little hole of an inch, and of an eliptical fhape, containing three eggs, pretty large, and of a fingular colour. *Obfervation made by M. Baillon, of Montreuil-fur-mer.*

† In Malta it is called *tolaride.*

the

the ftone curlews was referved in Malta for the mafter, before the introduction of our partridges, about the middle of laft century*.

Thefe birds do not, like the plovers, advance into the north in fummer; at leaft Linnæus has not inferted them in his *Fauna Suecica.* Willughby affures us, that in England they are found in Norfolk and in Cornwall; yet Charleton, who gives himfelf out for an experienced fportfman, confeffes that he never met with any. Indeed, the wild difpofition of the ftone curlew, and its retirement during the day, might long conceal it from the eyes of obfervers; and Belon, who firft difcovered it in France, remarks that no perfon could then tell him its name.

I had one of thefe birds a month or five weeks at my houfe in the country. It fed on foup, bread, and raw flefh, and preferred the laft. It ate not only in the day-time, but alfo at night; for its provifion, when given in the evening, was obferved to be diminifhed next morning.

This bird feemed to me of a peaceful temper, but timorous and wild; and I am induced to think that this is really the cafe, as it is feldom feen in the day, and prefers the obfcurity of night for its affociations. I obferved, that as foon as it perceived a perfon, even at a diftance, it endeavoured to efcape, and was fo overcome with fear, as to hurt itfelf in the flurry. It is, therefore, one of thofe birds which nature has

* In the time of the grand mafter, Martin de Redin.

deftined

deſtined to live remote from us, and has endued with the inſtinct for that purpoſe.

The one which I mention uttered no cry; it only made, two or three nights before its death, a ſort of feeble whiſtle, which was perhaps an expreſſion of pain, for the root of its bill and its feet were much gaſhed, owing to its violent ſtruggling in the cage, at the ſight of any new object.

[A] Specific character of the Thick-knee'd Buſtard, *Charadrius Oedicnemus:* "It is gray; its two primary wing-quills black, and in "the middle white; its bill ſharp; its feet cinereous." Theſe birds are very frequent in Norfolk; they breed in rabbit burrows; their eggs are olive, with reddiſh ſpots.

THE LONG LEGGED PLOVER

The LONG-SHANK.

L'Echasse *. *Buff.*

Charadrius Himantopus †. Linn. and Gmel.
Charadrius Autumnalis. Haffel.
Himantopus. Aldrov. Will. Sibb. Klein, &c.
Hæmantopus. Gefner.
The Long-legs. Ray and Sloane.
The Long-legged Plover. Penn. Lath. &c.

THE Long-Shank is among the birds what
the jerboa is among the quadrupeds : its
legs, which are thrice as long as the body, prefent a monftrous difproportion.   Such extrava-
gant and random productions of nature exhibit
the traces of her magnificent and boundlefs
plan ; and, like fhades in a picture, they heighten
by their contraft the beauty of the fcene.   The
enormous length of this bird's legs will hardly
allow it to reach the earth with its bill to gather

* *i. e.* Stilts.

† In Greek Ἱμαντοπυς, from ἱμας, a thong, and πυς, the foot ; fo
termed becaufe of its flender legs.   Pliny only writes the fame name
in Roman characters, *himantopus*, which the moderns have copied.
In Italian it is called, according to Belon, *Merlo Aquaiolo Grande*
(the great water black-bird) in Flemifh *Mathoen :* in German
Froembder Vogel, (the foreign bird) and alfo, according to Sibbald,
Dunn Bein and *Riemen Bein* (dun-fhank and thong-fhank).   In
Jamaica it has the appellation of *long-legged crane.*

its

its food; they are also slender, weak, and tottering *; and its three toes are disproportionally
small, and give a base too narrow for its tall
body. Hence the names of this bird in the ancient and modern languages refer to the softness
and pliancy of its legs, and to their extreme
length.

The slow laborious pace of this bird † seems
however to be compensated by its power of flying. Its wings are long, and extend beyond its
tail, which is pretty short; their colour, as well
as that of the back, is glossed with greenish blue;
the back of the head is brown-gray: the upper
side of the neck is mixed with blackish and
white: all the under surface is white from the
throat to the end of the tail: the feet are red,
and eight inches high, including the naked part,
which is three inches: the protuberance of the
knee is strongly marked on its smooth, slender
legs: the bill is black, cylindrical, a little flat
at the sides near the point, two inches and ten
lines long, close inserted on an elevated front,
which gives the head a round shape.

We are little acquainted with the habits of
this bird, whose species is scanty and rare ‡. It

* Aldrovandus, *tom.* iii. *p.* 444.

† *Incessus, nisi æquali alarum expansione librata sit, difficilis vide-
tur in tantâ crurum & pedum longitudine & exilitate.* Sibbald.

‡ We received a Long-Shank from Beauvoir, in Lower Poitou,
as an unknown bird; which proves that it seldom appears on those
coasts: this was killed in an old salt-pit; it was remarked in flying to stretch its legs backwards eight inches beyond the tail.

probably

probably lives on infects and worms, at the verge
of ftreams and marfhes. Pliny mentions it under
the name *himantopus*, and fays, " that it is a
" native of Egypt, and fubfifts chiefly on flies,
" and has never been kept more than a few
" days alive in Italy *." Yet Belon fpeaks of
it as an inhabitant of France ; and the Count
Marfigli faw it on the Danube. It appears alfo
to frequent the northern countries ; for though
Klein fays that he never faw it on the coafts of
the Baltic, Sibbald did in Scotland, and has accurately defcribed one that was killed near
Dumfries †.

The Long-Shank occurs alfo in the new
continent. Fernandez faw a fpecies or rather
a variety in New Spain ; and he fays that this
bird, an inhabitant of cold countries, does not
defcend till winter to Mexico; yet Sloane ranks
it among the Jamaica birds. It follows from
thefe apparently contradictory authorities, that
the fpecies of the Long-Shank, which contains
exceedingly few individuals, is diffufed or rather
difperfed, like that of the ring plover, in very
remote countries.—The Mexican Long-Shank,
indicated by Fernandez, is rather larger than
that of Europe ; it has a mixture of white in
the black of its wings : but thefe differences

* Lib. x. 46. Oppian likewife calls it *himantopus*. *(Exeutic,
lib.* ii.)

† *Scotia Illuftrata*, par. II. lib. iii. p. 19.

feem

feem infufficient to conftitute a diftinct fpecies *.

* *Comaltecatl.* Fernandez.
  Himantopus Mexicanus. Briff.

[A] Specific character of the Long-Shank, *Charadrius Himantopus :* " It is white, its back black ; its bill black, and longer than
" its head ; its feet red, and very long " This fingular bird is extremely rare in Britain. Sir Robert Sibbald gives a very full defcription of one fhot at a lake near the town of Dumfries; and Mr.
White has lately defcribed another fhot on the verge of Frinfham-
pond in Surrey : both of them have given engravings of the bird ;
but in Sibbald's figure the bill is reprefented quite ftraight, and in
White's it is fomewhat bent inwards.

THE PIED OYSTER CATCHER.

The OYSTER-CATCHER*.

L'HUITRIER; *vulgairement* LA PIE DE MER.
Buff.

Hæmatopus Ostralegus. Linn. and Gmel.
Scolopax Pica. Scopol.
Hæmatopus. Bel. Aldrov. Johnst. Sibb. &c.
Pica Marina. Charleton.
Ostralega. Briss.
The Sea Pie. Alb.
The Pied Oyster-catcher. Lath.

THOSE birds which disperse in our fields or retire under the shade of our forests, inhabit the most enchanting scenes, and the most peaceful retreats of nature. But such is not the lot of all the feathered race : some are confined to the solitary shores; to the naked beach, where the billows dispute the possession of the land; to the rocks, on which the surges dash and roar; and to the insulated shelving banks which are beaten by the murmuring waves. In these desert stations, so formidable to every other being,

* In French *Pie de Mer* (sea-pie) and sometimes *Becasse de Mer* (sea-woodcock.) In Gothland it is called *Marspitt* : in the isle of Oëland *Strandsk Jura* : in Norway *Tield, Glib, Strand-skuire, Strand-skade* : in the Feroe islands *Kielder* : in Iceland the male is named *Tilldur*, and the female *Tilldra*.

a few birds, fuch as the Oyfter-catcher, obtain
fubfiftance and fecurity, and even enjoy pleafure
and love. It lives on fea-worms, oyfters, limpets, and other bivalves, which it gathers on the
fand of the fea-fhore : it keeps conftantly on
the banks, which are left dry at low water, or on
the little channels, where it follows the refluent
tide ; and never retires farther than the fandy
hillocks which limit the beach. This bird has
alfo been called the *fea-pie*, not only on account
of its plumage, which is black and white, but
becaufe it makes a continual noife or cry, efpecially when it is in flocks : this cry is fhrill and
abrupt, repeated inceffantly whether at reft or on
the wing.

This bird is feldom feen on moft of the
French coafts ; yet it is known in Saintonge *,
and in Picardy †. In the latter province it
fometimes breeds, and arrives in very confiderable flocks when the wind is at eaft or north-
weft. They repofe on the fand of the beach,
waiting for a fair wind to waft them to their
ufual abode. They are believed to come from
Great Britain, where they are indeed very common, particularly on the eaftern coafts of that
ifland ‡. They alfo advance farther north ; for
they are found in Gothland, in the ifle of Oë-

* Belon.
† *Note communicated by M. Baillon, of Montreuil-fur-mer.*
‡ Willughby.

land,

land *, and in the Danifh iflands, as far as Norway † and Iceland.   On the other hand, Captain Cook faw them on the coafts of *Terra del
Fuego*, and near the Straits of Magellan ‡.   They
have been difcovered at Dufky Bay in New Zealand : Dampier remarked them on the fhores of
New Holland ; and Kæmpfer affures us, that
they are as common in Japan as in Europe.
Thus the Oyfter-catcher inhabits all the fhores
of the ancient continent ; and we need not be
furprized to find it in the new.   Father Feuillée
obferved it on the eaft of *Terra Firma* : Wafer
at Darien : Catefby at Carolina and the Bahama
Iflands : Page du Pratz at Louifiana §.   And
this fpecies, fo diffufed, has no variety, but every
where the fame, and disjoined from all others.
None indeed of the fhore birds has, with the ftature of the Oyfter-catcher, and its fhort legs,
the fame form of bill, or the fame habits and
œconomy.

* *Fauna Suecica*, Nº 161.
† Brunnich, *Ornithol. Borealis*, Nº 189.
‡ " Sea-pies or black Oyfter-catchers inhabit, with many other
" birds, the verge of thefe coafts, furrounded by immenfe floating
" beds of famphire on the eaft point of *Terra del Fuego*, and of the
" Strait."—*Cook's fecond Voyage.*
§ " The hatchet-bill is fo called, becaufe its bill is red, and fa-
" fhioned like the head of an hatchet : its feet are alfo of a very
" fine red, for which reafon it is often named *red-foot*.   As it lives
" wholly on fhell-fifh, it frequents the fea-fide, and is never feen
" on the land, except before fome great tempeft, which its retreat
" announces." Le Page du Pratz. *Hiftoire de la Louifiane, tom. ii.
p. 117.*

This

This bird is as large as the crow; its bill four inches long, contracted, and, as it were, compressed vertically above the noftrils, and flattened at the fides like a wedge as far as the tip, whofe fquare fection forms a cutting edge *: a ftructure peculiar to itfelf, and which enables its bill to raife up and detach the oyfters, limpets and other fhell-fifh from their beds and rocks.

The Oyfter-catcher is one of thofe few birds which have only three toes. This fingle circumftance has led fyftematic writers to range it next the buftard. Yet it is apparent how wide is the feparation in the order of nature; for not only this bird haunts the fea-fhores, it even fwims, though its feet are almoft entirely devoid of membranes. It is true that, according to M. Baillon, who obferved the Oyfter-catcher on the coafts of Picardy, its method of fwimming is merely paffive, but it has no averfion to repofe on the water, and leaves the fea for the land whenever it choofes.

Its black and white plumage, and its long bill, have given occafion to the inaccurate appellations of *fea-pie* and *fea-woodcock:* the name *Oyfter-catcher* is proper, fince it expreffes its mode of fubfifting. Catefby found oyfters in its ftomach, and Willughby entire limpets. The organ of digeftion is fpacious and mufcular, according to

* Le Page du Pratz, *ut fupra.*

Belon;

Belon; who adds, that the flesh of the bird is black and hard, and of a rank taste. Yet M. Baillon avers, that the Oyster-catcher is always fat in the winter, and that the young ones are tolerably pleasant food. He kept one more than two months in his garden, where it lived chiefly on earth-worms like the curlews, but also ate raw flesh and bread, with which it was well content. It drank indifferently either salt or fresh water, without shewing the least preference to either: yet in the state of nature these birds never inhabit the marshes or the mouths of rivers; they remain constantly on or near the beach; probably because fresh waters do not afford the proper subsistence.

The Oyster-catcher makes no nest; its eggs, which are grayish and spotted with black, it lays on the naked sand, out of the reach of the tide, but without any preparation for their reception; it only seems to select a high spot, strewed with fragments of shells. The number of eggs is usually four or five, and the time of incubation is twenty or twenty-one days: the female does not cover them assiduously; she seems, in this respect, to do as most of the birds that inhabit the sea-shores, and to leave the hatch to the sun's heat part of the day, leaving her charge at nine or ten o'clock of the morning, and not returning, unless on occasion of rain, till three in the afternoon. The young, when they have just burst the shell, are covered with a blackish

down.

down. They crawl on the fand the firft day; they foon begin to run, and then they conceal themfelves fo well in the tufts of grafs that it is difficult to find them *.

The bill and feet of the Oyfter-catcher are of a fine coral red: hence Belon has called it *hæmatopus* †, taking it for the *himantopus* of Pliny; but thefe two names ought not to be confounded, or applied to the fame bird. The former fignifies *red legs*, and might be referred to the Oyfter-catcher; but it is not found in Pliny, though Dalechamp reads it fo: and the latter, expreffive of tall, frail, and flender legs, belongs not to the Oyfter-catcher, but to the long-fhank. A word of Pliny in this paffage might have apprized Belon of his miftake; *præcipue ei pabulum mufcæ* ‡: the himantopus, which feeds on flies, can never be the Oyfter-catcher, which lives on fhell-fifh.

Willughby, in cautioning us not to confound this bird, under the name of *hæmantopus*, with the himantopus with long and delicate legs, feems to point out another miftake of Belon's; who, in his defcription of the Oyfter-catcher, attributes to it this delicacy of feet, incompatible, it would feem, with its mode of life, which obliges it to clamber on the edges of rocks. Its feet and

* *Note communicated by M. Baillon, of Montreuil-fur-mer.*

† From ημα, blood, and πυς, the foot.

‡ *i. e.* flies are its chief food, *lib.* x. 47.

*

toes

toes are covered with a rough and hard scale *.
It is more than probable, therefore, that here,
as in other cases, the confusion of names has
begotten that of the objects: the term *himantopus* ought to be appropriated to the long-
shank, and *hæmatopus* ought to be entirely ex-
punged, as vague and undefined.

The outer and middle toes of the Oyster-
catcher are connected as far as the first joint by
a portion of a membrane, and all the three are
surrounded with a membranous edge. Its eye-
lids are red, as well as its bill, and its iris is
gold-yellow; above each eye there is a little
white spot: the head, the neck, the shoulders,
are black, and also the upper surface of the
wings; but this black is deeper in the male
than in the female: under the throat there is a
white collar; all the under side of the body
from the breast is white, and the half of the tail,
whose tip is black: a white bar, formed by the
great coverts, intersects the dun black of the
wing; and these colours have probably sug-
gested the name of *pie*, though it differs from
that bird in every other respect, particularly in
the length of its tail, which is only four inches,
and covered three-fourths by the wing: the

* " The legs are strong and thick . . . and the feet remarkable
" by the hard and scaly skin which covers them. . . . Nature having
" not only given them a bill fashioned for opening oysters, but hav-
" ing also armed their legs and feet with scales against the sharp
" edges."—Catesby, *vol.* i. *p.* 85.

feet.

feet, together with the small naked part of the leg above the knee, are scarce two inches, though the bird is about sixteen inches long.

[A] Gmelin makes the Oyster-catcher a separate genus, containing only one species. It is a very shy bird, but common on most of the English coasts. If a person approaches its nest, it vents a shrill scream. These birds gather in large flocks during winter.

The SWIFT-RUNNER.

LE COURE-VITE. *Buff.*

Charadrius Gallicus. Gmel.
Curforius Europæus. Lath. Ind.
Pluvialis Morinellus Flavefcens. Gerin.
The Cream-coloured Plover. Lath. Syn.

THE two birds reprefented in N°s 795 and 892 of our *Pl. Enl.* are of a new genus, which required a diftinct name. They refemble the plover in the feet, which have only three toes, but differ in the fhape of the bill, which is curved, whereas that of the plovers is ftraight and inflated near the end. The firft was killed in France, where it had probably ftrayed, fince no other has been feen. It ran with rapidity along the fhore, and hence it derived its name. We have fince received from the coaft of Coromandel a bird entirely fimilar in form, and different only in its colours; fo that it may be regarded as a variety, or at leaft a kindred fpecies. Both of them have larger legs than the plovers; they are as large, but not fo thick; their toes are very fhort, particularly the two lateral ones. The plumage of the firft is gray, wafhed with rufous brown; on the eye

is

is a lighter ftreak, almoft white, which ftretches
backwards, and below it a black ftreak rifing
from the outer angle of the eye : the top of the
head is rufous ; the quills of the wing are black,
and each feather of the tail, except the two
middle ones, has a black fpot, with another white
one near the point.

The fecond, which came from Coromandel,
is rather fmaller than the firft ; the fore fide of
the neck and breaft of a fine chefnut-rufous,
which lofes itfelf in black on the belly ; the
quills of the wing are black ; the upper furface
gray ; the lower belly white ; the head enveloped with rufous nearly as in the firft : in both
the bill and feet are yellowifh white.

THE TURNSTONE.

The TURN-STONE.

LE TOURNE-PIERRE. *Buff.*

Tringa-Interpres. Linn. and Gmel.
Arenaria. Briff.
Morinellus Marinus. Will. and Ray.
The Hebridal Sandpiper. Penn. and Lightfoot.
The Turn-ftone, or Sea Dotterel. Edw. Lath. &c.

WE adopt the name *Turn-ftone*, given by Catefby, becaufe it indicates the fingular habit which this bird has of turning over the ftones at the water's-edge, to difcover the worms and infects that lurk under thefe; whereas all the other fhore-birds content themfelves with fearching in the fand or mud. " Being at fea," fays Catefby, " forty leagues from Florida, in " the latitude of 31 degrees, a bird flew on our " fhip, and was caught. It was very alert in " turning the ftones that lay befide it: in doing " fo, it ufed only its upper mandible, and was " able to turn over very nimbly ftones of three " pounds weight *." This action implies fingular force and dexterity in a bird which is hardly fo large as the dufky fandpiper.

* Carolina, vol. i. p. 72.

It

It has a bill of a harder and more horny fubftance than the other little fhore-birds, and it
forms a fmall family amid that numerous tribe:
the bill is thick at the root, and gradually tapers
to a point: the upper-part is fomewhat compreffed, and appears to rife with a flight curve:
it is black, and an inch long: the feet have no
membranes, are pretty fhort, and of an orange-
colour.

The plumage of the Turn-ftone refembles
that of the ringed plover, in the white and black
which interfect it, but without tracing diftinctly
a collar, and in the mixture of rufous on the
back. This fimilarity in its colours has probably mifled Brown, Willughby, and Ray, who have
given it the appellation *morinellus* (dotterel);
though it is of a kind entirely diftinct from the
plovers, being furnifhed with a fourth toe and a
differently fhaped bill.

The fpecies of the Turn-ftone is common to
both continents. It is known on the weft coaft
of England, where thefe appear generally in fmall
companies of three or four *. They are equally
known on the coaft of Norfolk †, and in fome of
the iflands of Gothland ‡; and we have reafon
to believe that it is the fame bird which, on the
fhores of Picardy, is called the *bune*. We
received one from the Cape of Good Hope,

* Willughby.
† *Id. Ibid.*
‡ *Heligholmen* and *Clafen.* Fauna Suecica, Nº 154.

which

which was of the fame fize, and, except fome
flight differences, of the fame colour with that
of Europe. Catefby faw thefe birds near the
coafts of Florida * ; and we cannot divine why
Briffon reckons the American Turn-ftone different from the European. We received one alfo
from Cayenne, which was only fomewhat larger;
and Edwards mentions another fent from Hudfon's Bay. Thus this fpecies, though it contains few individuals, has, like many other aquatic birds, fpread from north to fouth in both
continents, following the fea-fhore, which yields
it fubfiftence.

The gray Turn-ftone of Cayenne appears to be
a variety of this fpecies, and to which we fhould
refer the two birds reprefented in the *Planches
Enluminées,* under the appellations of *Coulon-chaud
de Cayenne,* and *Coulon-chaud gris de Cayenne*;
for we can perceive no effential difference. We
fhould even regard them as the females of the
firft fpecies, of which the male would have
ftronger colours; but we fufpend our judgment,
becaufe Willughby affures us that he could dif-
cern no difference between the plumage of the
male and female of the Turn-ftones which he
defcribed.

* " Comparing this bird with the defcription which Mr. Wil-
" lughby gives of the fea-lark (turn-ftone), I found it was the fame
" fpecies." Catefby.

[A] Specific character of the Turn-ftone. *Tringa-Interpres:* " Its
" feet are red, its body black, variegated with white and ferrugi-
" nous; its breaft and belly white."

The WATER OUZEL*.

LE MERLE D'EAU. *Buff.*

Sturnus Cinclus. Linn. and Gmel.
Merula Aquatica. Gefner, Ald. Johnft. Briff. &c.
Turdus Aquaticus. Klein.
Motacilla-Cinclus. Scop.
Turdus-Cinclus. Lath.

THOUGH this bird has received the names of Ouzel, of Stare, of Thrufh, or of Wag-tail, it is a water-fowl, which frequents the lakes and rills on lofty mountains. It refembles the blackbird in fize, only fomewhat fhorter, and in the colour of its plumage, which is almoft black; it has alfo a white fpace on the breaft: but it is taciturn, it walks leifurely with meafured fteps, and runs befide the fprings and brooks, which it never leaves †; preferring the limpid ftreams, whofe fall is rapid, and whofe bed is broken with ftones and fragments of rocks. It is found near torrents and cafcades,

---

* In Italian *Merla Aquaiola:* near Belinzone *Lerllchtrollo,* and about lake Maggiore *Folun d'Aqua,* according to Gefner: in German *Bach-Amfel, Waffer Amfel:* in Swifs *Waffer Treftle:* in Swedifh *Watn-Stare.*

† Schwenckfield.

and

THE WATER OUZEL.

and efpecially in the pebbly channels of clear
rivulets *.

Its habits are very fingular. Aquatic birds
with palmated feet fwim or dive; thofe which
inhabit the fhores, without wetting their body,
wade with their tall legs; but the Water-
Ouzel walks quite into the flood, following
the declivity of the ground: it is obferved to
enter by degrees, till the water reaches its neck;
and it ftill advances, holding its head not higher
than ufual, though completely immerfed: it
continues to walk under the water, and even
defcends to the bottom, where it faunters as on
the dry bank. We are indebted to M. Hebert
for the firft account of this extraordinary habit,
which I know not to belong to any other bird.
I fhall here give the obfervations which he was
pleafed to communicate to me.

" I lay ambufhed on the verge of the lake
" Nantua, in a hut formed of pine-branches
" and fnow, where I patiently waited till a boat,
" which was rowing on the lake, fhould drive
" fome wild-ducks to the water's edge. I ob-
" ferved without being perceived: before me
" was a fmall inlet, the bottom of which gently

---

* The Water-Ouzel has a very wide mouth; its feathers are
greafy like the ducks, which enables it the eafier to dive under water
for aquatic infects: it forms its neft with mofs near rivulets, and fa-
fhions it like a vault: its eggs are four in number. *Extract of a letter from Dr. Hermann to M. de Montbeillard, dated Strafburg, 22d September* 1774.

" fhelved,

" shelved, and might be about two or three feet
" deep in the middle. A Water-Ouzel stopped
" here more than an hour, and I had full leisure
" to view its manœuvres. It entered into the
" water, disappeared, and again emerged on the
" other side of the inlet, which it thus repeat-
" edly forded. It traversed the whole of the bot-
" tom, and seemed not to have changed its ele-
" ment, and discovered no hesitation or reluc-
" tance in the immersion. However, I per-
" ceived several times, that as often as it waded
" deeper than the knee, it displayed its wings,
" and allowed them to hang to the ground. I
" remarked, too, that when I could discern it at
" the bottom of the water, it appeared inveloped
" with air, which gave it a brilliant surface; like
" some forts of beetles, which are always in wa-
" ter, inclosed with a bubble of air. Its view
" in dropping its wings on entering the water,
" might be to confine this air; it was certainly
" never without some, and it seemed to quiver.
" These singular habits of the Water-Ouzel
" were unknown to all the sportsmen whom I
" have talked with; and, perhaps, without the
" accident of the snow-hut, I should have ever
" been ignorant of them: but I can aver, that
" the bird came to my very feet, and, that I
" might observe it, I did not kill it *."

The history of birds presents few facts more
curious than the foregoing. Linnæus had

* *Note communicated by M. Hebert to M. le Comte de Buffon.*

rightly

rightly faid, that the Water-Ouzel defcends
into the ftreams, and again emerges with much
dexterity * ; and Willughby had remarked that,
though cloven-footed, this bird dived: but neither of them feems to have known that it
plunged in order to walk on the bottom. We
may eafily fuppofe, that for this purpofe a pebbly channel and clear water are requifite, and
that a flimy ground would be altogether improper. Accordingly, this bird is found only in
mountainous countries, at fources of rivers, and
in the torrents which pour down from the
rocks; as in Weftmorland and other hilly parts
of England † : in France, among the mountains
of Bugey and Vofges, and in Switzerland ‡. It
fits on the ftones through which the rills wind
their courfe. It flies very fwiftly ftraight forward, razing the furface of the water, like a
kingfifher. When on wing, it utters a feeble cry, efpecially in the feafon of fpring: it
then affociates with its female, though at all
other times it goes fingle ‖. The female lays
four or five eggs; conceals her neft with great
care, and often places it near the paper-mills
conftructed on brooks.

The feafon in which Hebert faw the Water-
Ouzel, proves that it is not a bird of paffage.

* *Fauna Suecica.*
† Willughby.
‡ *Idem.*
‖ *Idem.*

It remains all winter in our mountains, and
dreads not the rigour of winter even in Sweden,
where it feeks the cataracts and whirlpools,
which are not affected by the frost *.

The nails of this bird are very strong and
curved, and ferve to clasp the pebbles, as it
walks at the bottom of the water : the feet have
the fame conformation as thofe of the land
ouzels : like them alfo, it has the hind toe and
nail ftronger than thofe placed before, and thofe
toes are diftinctly parted, and without any portion of membrane, as Willughby fuppofed. The
leg is feathered to the knee ; the bill is fhort and
flender, both mandibles tapering equally, and
arched flightly near the point.

The bill of this bird, the feet and the neck
being fhort, we might conceive it requifite to
walk under the water, to catch the fmall fifh
and aquatic infects on which it feeds. Its plumage, which is thick and clothed with down,
feems impenetrable by water, which enables it
to remain without inconvenience in that fluid;
its eyes are large, of a fine brown ; the eye-lids
white, and they muft be kept open under water,
that the bird may diftinguifh its prey.

A fine white fpace covers the neck and breaft;
the head and the upper fide of the neck, as far
as the fhoulders and the border of the white
fpace, are rufty afh-colour, or chefnut; the

* *Fauna Suecica.*

back,

back, the belly, and the wings, which reach
not beyond the tail, are of a blackifh and flaty
cinereous; the tail is very fhort, and has nothing
remarkable.

[A] Specific character of the Water-Ouzel, *Turdus-Cinclus* :
" Its bill and feet are black; its ftraps white; its tail and rump
" gray and brown."

The WATER-THRUSH.

LA GRIVE D'EAU. *Buff.*

Tringa Macularia. Linn. and Gmel.
Turdus Aquaticus. Briff.
The Spotted Tringa. Edw.
The Spotted Sandpiper. Penn. and Lath.

THIS bird has the fpeckled plumage and the fize of the throftle : its feet refemble thofe of the preceding ; its nails are large and hooked, particularly the hind one ; but its bill is fimilar to that of the purre, of the dufky fandpiper, and of other fmall fhore birds, and the lower part of its leg is naked : it is therefore wide of being a thrufh. It appears to be a foreign fpecies, little related to the European birds : yet Edwards prefumes that it is common to both continents, as he received one from the county of Effex, where it had ftrayed, no other having ever been feen there.

The bill is eleven or twelve lines in length : it is flefh-coloured at the bafe, and brown near the point ; the upper mandible is marked on each fide by a furrow, which extends from the noftrils to the extremity of the bill ; the upper furface of the body is fprinkled with blackifh fpots

on

on an olive-brown ground, as the under furface is fpeckled on a lighter and whitifh ground; there is another white bar above each eye, and the quills of the wing are blackifh; a fmall membrane joins the outer toe, near the root, to the middle one.

[A] Specific charaſter of the Spotted Sandpiper, *Tringa Maculata:* " The bafe of its bill and its feet are carnation; its body " fpotted all over; its eye-brows, and a double bar on its wings, " white."

The KNOT.

LE CANUT. *Buff.*

Tringa-Canutus. Linn. and Gmel.

IT is probable that in fome of the northern countries there are traditionary anecdotes of this bird, fince it retains the name of Canute, the Dane, king of England *. It would much refemble the gray lapwing, were it as large, and its bill otherwife fhaped. This is pretty thick at the bafe, and tapers gradually to the extremity, which is not very pointed, yet not inflated like the bill of the lapwing: all the upper fide of the body is cinereous and waved; the white tips of the great coverts trace a line on the wing; blackifh crefcents on the white-gray ground mark the feathers of the rump: all the under fide of the body is white fpotted with gray on the throat and breaft; the lower part of the leg is naked; the tail does not exceed the clofed wings. Willughby fays, that he faw one of thefe birds in Lincolnfhire, about the begin-ning of winter, and that they remain there two

* That monarch is faid to have been remarkably fond of the flefh of this bird. *Willughby.*

or

or three months, in flocks by the fea-fhore, and
afterwards difappear: he adds, that he faw
another near Liverpool. That which Edwards
defcribes, was found in the London market, in
the hard winter of 1740; which makes me
think that they never advance to the fouth of
Great Britain, but in the moft fevere feafons.
They muft, however, be more common in the
northern parts of that ifland, fince Willughby
defcribes a method of fattening them, by feeding
them with bread foaked in milk, and fpeaks of
the exquifite flavour thus communicated to their
flefh. He fubjoins, that, at firft fight, a perfon
would not diftinguifh them from the fandpipers
(tringæ), but by the white bar on the wing.
The bill, he remarks, too, is of a harder fubftance
than ufual in other birds, in which its ftructure
refembles that of the woodcock.

An indication given by Linnæus would fhew,
that this bird is found in Sweden; yet does a
difficulty occur: for, according to Willughby,
the feet of the *Knot* are cloven, but Linnæus reprefents his *Canutus* as having its outer toe connected by the firft joint to that of the middle.
If both thefe obfervers have been accurate, we
muft regard the two birds as belonging to diftinct fpecies.

[A] Specific character of the Knot, *Tringa-Canutus:* " Its bill
" is fmooth, its feet afhy, its primary wing-quills ferrated, its
" outermoft tail-quill fpotlefs white."

The RAILS.

LE RALES. *Buff.*

THESE birds conftitute a large family, and their habits are different from thofe of the other fhore-birds, which refide on fands and gravel. The Rails, on the contrary, inhabit only the flimy margins of pools and rivers, efpecially low grounds covered with flags and other large marfh plants. This mode of living is habitual and common to all the fpecies of water rails. The land rail frequents meadows, and from the difagreeable cry, or rather rattling in the throat, of this bird, is derived the generic name *. In all the Rails, the body is flender, and fhrunk at the fides; the tail extremely fhort; the head fmall; the bill pretty like that of the gallinaceous kind, though much longer, and not fo thick; a portion of the leg above the knee is bare; the three fore toes without membranes, and very long: they do not, like other birds, draw their feet under their belly in flying, but allow them to hang down; their wings are fmall and very concave, and their flight is fhort.

---

* *Râler*, in French, fignifies *to rattle in one's throat*.

THE LAND RAIL.

The LAND RAIL.

LE RALE DE TERRE, *ou* GENET. *Buff.*

FIRST SPECIES.

Rallus-Crex. Linn. and Gmel.
Gallinula-Crex. Lath. Ind.
Ortygometra. Gesner, Aldrov. Will. Johnst. &c. *
Crex. Gesner, Aldrov. Charleton, &c.
Rallus Genistarum. Briss.
Rallus Terrestris. Klein.
The Land Hen. Will.
The Daker Hen, or Rail. Alb.
The Corn Crex. Sibb. Scot.
The Corn Craker. Martin's West. Isles.
The Crake Gallinule. Lath. Syn.

IN wet meadows, from the time the grass is grown till it be cut down, there issues from the thickest part of the herbage a raucous voice, or rather a broken, harsh cry, *crĕk, crĕk, crĕk,* much like the noise made by stripping forcibly the teeth of a large comb under the fingers: as we approach the sound retires, and is heard at a

* It is denominated in Greek, Italian, French and German, *the mother or king of the Quails,* Ορτυγομητρα, *Re delle Quaglie, Roi des Cailles, Wachtel Koenig:* in German it is also named, from its cry, *Schryck, Schrye:* in Silesia *Snercker:* in Poland *Chrosciel, Derkacz, Kasper:* in Sweden *Korn Knarren*; and in the province of Upland *Aengsnaerpa:* in Norwegian *Akerrire, Ager-hone.* This bird lurks frequently in broom, and hence the name it sometimes has in French, *Genet,* and the appellation which Brisson gives it, *Rallus Genistarum.*

remove

remove of fifty paces. It is the Land Rail that emits this cry, which might be taken for the croaking of a reptile *. This bird seldom escapes by flight, but almost always walks nimbly through the thickest grass, where it leaves a remarkable track. It begins to be heard about the 10th or 12th of May, at the same time with the quails, which it seems ever to accompany †. Hence, as the quails too inhabit meadows, and as the Land Rail is less common, and rather larger, it has been supposed to be their leader ‡, and therefore called the *king of the quails*. But it differs from these in the characters of its conformation, which are common to the other rails, and in general to the marsh birds §, as Aristotle has well observed ‖. The chief resemblance which this rail bears to the quail consists in its plumage, which however is browner and more golden; fulvous predominates on the wings; blackish and rusty form the colours of the body; these are disposed on the flanks by transverse lines, and are all paler in the female, which is rather smaller also than the male.

A gratuitous extension of the imaginary analogy between the Land Rail and the quail has likewise impressed the notion of an equal fecun-

* Longolius, *apud Gesnerum.*
† They arrive and retire together, according to Longolius.
‡ Aristotle, *Hist. Animal.* lib. viii. 12.
§ Klein.
‖ Lib. viii. 2.

dity.

dity. But, from repeated obfervations, we are
affured, that it feldom lays eight or ten eggs,
never eighteen or twenty, as fuppofed : indeed,
were we to admit fuch multiplication, the fpecies would be more numerous, confidering how
well the neft is concealed. This neft is negli-
gently conftructed with a little mofs or dry
grafs, and placed ufually in fome fmall hollow
in the turf: the eggs, larger than thofe of the
quail, are fprinkled with broader reddifh fpots.
The young crakes run as foon as they burft the
fhell, following their mother, but quit not the
meadow till the fcythe fweeps away their habitation. The late hatches are plundered by the
hands of the mower. All the other broods
then fhelter themfelves in the fields of buck-
wheat, among oats, and in wafte grounds overfpread with broom, where they are found often
in fummer: a few return again to the meadows
about the end of that feafon.

We may know when a dog lights on the fcent
of the Land Rail, from his keen fearch, his
number of falfe tracks, and the obftinacy with
which the bird perfifts in keeping the ground,
infomuch that it may be fometimes caught by
the hand: it often ftops fhort, and fquats down;
the dog pufhing eagerly forward, overfhoots the
fpot, and lofes the trace; the Rail, it is faid, pro-
fits by his blunder, and runs back on its path;
nor does it fpring till driven to the laft extremity, and then it rifes to a good height before

it

it ftretches onwards. It flies heavily, and never to a great diftance. It is ufually feen to alight, but in vain fhould we fearch for it; before the fowler has reached the fpot, the bird has tripped off more than an hundred paces. The fleetnefs of its feet compenfates for the tardinefs of its wings: all its little excurfions, its windings, and its doublings in the fields and meadows, are performed by running. When about however to retire into other countries, it feels, like the quail, unufual vigour, which fits it for perform- ing the diftant journey *. It commences its flight during the night, and aided by a favoura- ble wind it advances into the fouth of France, where it attempts the paffage of the Mediterra- nean. Many perifh, no doubt, in thefe migra- tions, and it is remarked that their numbers are fewer on their return.

The Land Rail is never feen in the fouth of France but in its paffage: it does not breed in Provence †. Belon fays, that it is rare in Can- dia, though pretty common in Greece and Italy: it is found therefore in that ifland only in its tranfits of fpring and autumn. The migrations of this bird extend more to the north than to

* I afked the Tatares how this bird, not being able to fly, could retire in winter: they told me that the Tatares and the Affanians knew well that it could not of itfelf pafs into another country; but that when the cranes retire in autumn, each takes a rail upon its back, and conveys it to a warmer climate.

Gmelin, *Voyage en Siberie*, tom. ii. *p.* 115.

† *Memoirs communicated by the Marquis de Piolene.*

the

the fouth; and, notwithftanding the flownefs of
its flight, it penetrates into Poland *, Sweden †,
Denmark, and even Norway ‡. It is faid to be
rare in England §, and found only in fome particular diftricts, though common in Ireland ‖.
Its motions feem to obferve the fame order in
Afia as in Europe; and in Kamtfchatka the
month of May is likewife the term of their
arrival ¶.

The Land Rail repairs to the northern countries as much for the fake of cool fituations, as
to obtain its proper food; fince, though it eats
feeds, efpecially thofe of broom, trefoil, grom-
well, and fattens in the cage on millet and
grain **, it prefers infects, flugs, worms; and
thefe, which are neceffary for its young, can be
found only in fhady wet grounds ††. But when
grown up, every fort of aliment fuits it: it becomes fat, and its flefh exquifite. It is caught,
like the quail, by a net, into which it is decoyed
by an imitation of its cry, *crĕk, crĕk, crĕk,* by

* Rzyczyfki.
† Linnæus.
‡ Muller, Brunnich.
§ Turner fays, that he never faw or heard it anywhere but in
Northumberland: yet Dr. Tancred Robinfon avers that it is found
alfo in the northern part of Great Britain; and Sir Robert Sibbald
reckons it among the birds of Scotland.
‖ Willughby and Ray.
¶ In the Kamtfchadale language, the month of May is called
Tava Koatch, which fignifies the month of quails.
** Aldrovandus.
†† Willughby, Schwenckfeld, Linnæus.

rubbing

rubbing hard the blade of a knife on an indented bone *.

Moſt of the names given in different languages to this bird are evidently formed to imitate this ſingular cry †.   Hence Turner, and
ſome other naturaliſts, have ſuppoſed it to be
crex of the ancients.   But that term appears to
have been applied by the ancients to other birds.
Philus gives the *crex* the epithet of βραδυπ]εϱος,
or *ſluggiſh-winged*, which would indeed ſuit the
Land Rail.   Ariſtophanes repreſents it as mi-
grating from Libya :   Ariſtotle ſays, that it is
quarrelſome, which may have been attributed to
it from the analogy to the quail ; but he adds,
that the *crex* ſeeks to deſtroy the neſts of the
black - bird ‡, which cannot apply to the rail,
ſince it never inhabits the woods.   Still leſs is
the *crex* of Herodotus a rail, for he compares
its ſize to that of the ibis, which is ten times
larger §.   The avoſet, too, and the teal, have
ſometimes the cry *crex, crex :* and the bird
which Belon heard repeating that cry on the
banks of the Nile, is, according to his account,
a ſpecies of godwit.   Thus the ſound repre-
ſented by the word *crex*, belonging to ſeveral
ſpecies, is not ſufficiently preciſe to diſtinguiſh
the Land Rail.

* Longolius.
† *Schryck, Scherck, Korn-Knaerr, Corn-Crek, &c.*
‡ Lib. ix. 1.
§ See the Article of the *Ibis*.

[A] Specific

[A] Specific character of the Land Rail, *Rallus-Crex*: " Its
" wings are rufous ferruginous." This bird leaves our island in
winter: on its first arrival it weighs only fix ounces, but fattens
fo much during its ftay as to weigh eight ounces before it retires.
The Land Rails appear numerous in the ifle of Anglefea, about the
end of May, and are fuppofed to pafs from thence into Ireland,
where the humid face of the country is fo congenial to their nature.

The WATER RAIL.

SECOND SPECIES.

Rallus Aquaticus. Linn. Gmel. Briff. &c.
The Water Rail, Bilcock, or Brook Ouzel*. Will.

THE Water Rail runs befide ftagnate water as fwiftly as the land rail through the fields. It alfo lurks conftantly among the tall herbs and rufhes. It never comes out but to crofs the water by fwimming or running; for it often trips nimbly along the broad leaves of the water-lily which cover pools †. It makes fmall tracks over the tall grafs; and as it always keeps the fame paths, it may be eafily caught by noofes fet in them ‡. Formerly, the fparrow-hawk or falcon § was flown at it; and in that fport the greateft difficulty was to fet up the

---

* In German *Schwartz Waffer Heunle* (black water-hen) *Aefch-heunlin* (cinereous fowl). Gefner gives it the name *Samet-hunle,* or velvet hen, on account of its foft plumage. At Venice it is called *Forzane* or *Porzana,* which appellation is alfo beftowed on the water hens. In Denmark it is denominated *Vagtel-Konge :* in Norway *Band-rire, Strand-fnarre, Vand-hone, Vand-vagtel :* and in the Feroe iflands *Jord-koene.*

† Klein.
‡ Belon.
§ Belon and Gefner.

§

bird,

THE WATER RAIL.

bird, for it ftuck to its concealment with the obftinacy of the land rail. It caufes the fame trouble to the fportfman, raifes the fame impatience in the dog, which it mifleads and diftracts, and protracts as long as poffible its fpringing. It is nearly as large as the land rail, but its bill is longer, and reddifh at the point; its feet are of a dull red: Ray fays, that in fome fpecies thefe are yellow, and that this difference may proceed from the fex. The belly and fides are ftriped acrofs with whitifh bars on a blackifh ground: the colours are difpofed the fame as in all the rails: the throat, the breaft, the ftomach, are of a fine flate-gray: the upper furface is of an olive brown rufous.

Water Rails are feen near the perennial fountains during the greateft part of the winter: yet, like the land rails, they have their regular migrations. They pafs Malta in the fpring and autumn *. The Vifcount de Querhoënt faw fome fifty leagues off the coafts of Portugal on the 17th of April; they were fo fatigued that they fuffered themfelves to be caught by the hand †. Gmelin found thefe birds in the countries watered by the Don. Belon calls them

* Note communicated by M. Defmazy.

† " I tried," fays M. de Querhoënt, " to raife fome : they " thrived wonderfully at firft, but after a fortnight's confinement " their long legs grew paralytic, and the birds could only crawl on " their knees; at laft they expired." Gefner fays, that haying long fed one, he found it to be peevifh and quarrelfome.

black

black rails, and fays they are *every where known,* and that the fpecies is more numerous than the *red rail* or land rail.

The flefh of the Water Rail is not fo delicate as that of the land rail, and has even a marfhy tafte, nearly like that of the gallinule.

[A] Specific character of the Water Rail: " Its wings are " gray, fpotted with brown ; its flanks fpotted with white; its bill " fulvous below." It continues the whole year in England.

The MAROUETTE.

THIRD SPECIES.

Rallus-Porzana. Linn. and Gmel.
Gallinula-Porzana. Lath. Ind.
Gallinula Ochra. Gesner.
Porcellana, Porzana, Grugnetto. Aldr.
Rallus Aquaticus Minor, five Marouetta. Briss.
The Spotted Water-hen. Penn.
The Spotted Gallinule. Lath. Syn.*

THIS is a small water rail, not exceeding a
lark in size. All the ground of its plumage is olive-brown, spotted and clouded with
whitish, whose lustre gives this dark shade an
enamelled glofs; whence it has been called the
pearled rail. Frisch denominates it improperly
the *spotted water-hen.* It appears at the same
seafon with the great water rail: it haunts
marshy pools: it lurks and breeds among the
reeds: its nest is fashioned after the manner of
a gondola, and compofed of rushes interwoven
and fastened at the ends to the stalk of a reed;
fo that, like a small boat or cradle, it rifes and

* In Picardy it is called *Girardine,* and in the Milanefe *Girar-
dina:* in fome parts of France *Cocouan,* according to Brisson: in
the Bolognefe *Porzana:* and in Alface *Winkernell,* according to
Gesner.

finks

finks with the water. It lays feven or eight eggs; and the floating young are hatched all black. Their education is fpeedy; for they run, fwim, dive, and foon feparate, each to lead a folitary, favage life, which prevails even in the feafon of love: fince, except during actual coition, the male difcards his female, pays no tender attentions or careffes, indulges in no frolics or joyous airs, and feels none of thofe foft delights, the fweet preludes of fruition. Unhappy beings, who never breathe a figh to the objects of their paffion! infipid loves, whofe fole end is to procure pofterity!

Its habits wild, its inftinct ftupid, the *Marouette* is unfufceptible of education, nor is even capable of being tamed. We raifed one, however, which lived a whole fummer on crumbs of bread and hemp-feed: when by itfelf, it kept conftantly in a large bowl of water; but if a perfon entered the clofet where it was fhut, it ran to conceal itfelf in a fmall dark corner, without venting cries or murmurs. In the ftate of liberty, however, it has a fharp piercing voice, much like the fcream of a young bird of prey: and though it has no propenfity to fociety, as foon as one cries, another repeats the found, which is thus conveyed through all the reft in the diftrict.

The *Marouette*, like all the rails, is fo obftinately averfe to rife, that the fportfman often feizes it with his hand, or fells it with a ftick.

If

If it finds a bufh in its retreat, it climbs upon it, and from the top of its afylum beholds the dogs -brufhing along in fault: this habit is common to it and to the water rail. It dives, fwims, and even fwims under water, when hard pufhed.

Thefe birds difappear in the depth of winter, but return early in the fpring; and even in the month of February they are common in fome provinces of France and Italy. Their flefh is delicate and much efteemed; thofe in particular which are caught in the rice-fields in Piemont are very fat, and of an exquifite flavour.

[A] Specific character of the *Rallus-Porzana*: " Its two middle " tail-quills are edged with white; its bill and feet fomewhat " olive."

FOREIGN BIRDS of the ANCIENT CONTINENT,

WHICH ARE RELATED TO THE RAIL.

———

The TIKLIN, or PHILIPPINE RAIL.

FIRST SPECIES.

Rallus Philippensis. Linn. Gmel. and Briff.

THERE are four different species known by
the name of *Tiklin* in the Philippine islands.
The present is remarkable for the neatness and
agreeable contrast of its colours: a gray space
covers the fore side of the neck ; another space
of chesnut rufous covers the upper side of it and
the head ; a white line extends above the eye ; all
the under side of the body is enamelled as it were
with little cross lines, alternately black and white
in festoons ; the upper surface is brown, clouded
with rusty, and sprinkled with small white drops
on the shoulders and the edge of the wings, of
which the quills are intermixed with black, white,
and chesnut. This bird is rather larger than the
water rail.

[A] Specific character of the *Rallus Philippensis :* " It is brown,
" below striped with gray ; its eye-brows white ; its neck tawny
" below."

The BROWN TIKLIN.

SECOND SPECIES.

Rallus Fuscus. Linn. Gmel. and Briff.
The Brown Rail. Lath.

THE plumage of this bird is of an uniform
dull brown, only wafhed on the throat
and breaft with a purple vinous tint, and broken
under the tail by a little black and white on
the lower coverts. It is as fmall as the preceding.

[A] Specific charaƈter of the *Rallus Fufcus :* " It is brown, its
" vent waved with white, its feet bright yellow."

The STRIPED TIKLIN.

THIRD SPECIES.

Rallus Striatus. Linn. and Gmel.
Rallus Philippensis Striatus. Briff.

THIS is of the fame fize with the preceding.
The ground of its plumage is fulvous
brown, croffed, and, as it were, worked with
white lines; the upper part of the head and neck
is chefnut-brown : the ftomach, the breaft, and
the neck are olive-gray; and the throat is rufty
white.

[A] Specific character of the *Rallus Striatus :* " It is blackifh
" waved with white, its throat tawny."

The COLLARED TIKLIN.

FOURTH SPECIES.

Rallus Torquatus. Linn. and Gmel.
Rallus Philippensis Torquatus. Briss.
The Banded Rail. Lath.

THIS is rather larger than the land rail. Its upper surface is brown, tinged with dull olive; its cheeks and throat are foot-colour; a white track rises from the corner of the bill, passes under the eye, and extends behind; the fore side of the neck, the breast, the belly, are blackish‑brown, striped with white lines; a band of fine chesnut of the breadth of the finger, forms a half collar above the breast.

[A] Specific character of the *Rallus Torquatus :* " It is brown, " below waved with white, a white line below its eyes."

FOREIGN BIRDS of the NEW CONTINENT,

WHICH ARE RELATED TO THE RAIL.

The LONG-BILLED RAIL.

FIRST SPECIES.

Rallus Longirostris. Gmel.

THE species of the rails are more diversified, and perhaps more numerous, in the deluged and swampy grounds of the new, than in the dryer countries of the ancient continent. It appears that two of these are smaller than the rest, and that the present is, on the contrary, larger than any of the European. Its bill also is longer, even than in proportion; its plumage is gray, or a little rusty on the fore side of the body, and mixed with blackish or brown on the back and the wings; the belly is striped with white and black cross bars, as in most of the other rails. Two species, or at least two varieties of these birds, are found in Cayenne; and they differ widely in size, some being as large as a godwit, and others scarcely equal to the common water rail.

The K I O L O.

SECOND SPECIES.

Rallus Cayanenſis. Gmel.
The Cayenne Rail. Lath.

THIS is the name by which the natives of Cayenne expreſs the cry or puling of this Rail. It is heard in the evening at the ſame hour with the *tinamous*, that is, at ſix o'clock, the inſtant the ſun ſets in the equatorial climates. Their cry is the ſignal to aſſemble ; for in the day-time they lurk diſperſed and ſolitary in the wet buſhes. They make their neſt in the little low branches, and it conſiſts of a ſingle ſort of reddiſh herb ; it is raiſed into a ſmall vault to prevent the rain from penetrating. This Rail is rather ſmaller than the *marouette* ; the fore ſide of its body and the crown of its head are of a fine rufous, and the upper ſurface is waſhed with olive-green on a brown ground. We conceive that Edwards's Penſylvanian rail is the ſame with this *.

---

* *Rallus Virginianus.* Linn. and Gmel.
Rallus Aquaticus, var. 1. Lath.
Rallus Penſylvanicus. Briſſ.
The American Water-Rail. Edw.
The Virginian Rail. Penn.

" Above it is brown, below tawny-brown ; its tail-quills brown ;
" its eye-brows and its throat white." *Latham.*

x

The SPOTTED RAIL of CAYENNE.

THIRD SPECIES.

Rallus Variegatus. Gmel.
The Variegated Rail. Lath.

THIS handsome Rail, which is one of the largest, has brown-rufous wings; the rest of the plumage spotted, streaked and edged with white, on a jet ground. It is found, too, in Guiana.

The VIRGINIAN RAIL.

FOURTH SPECIES.

Rallus Carolinus. Linn. and Gmel.
Gallinula Carolina. Lath. Ind.
Rallus Terrestris Americanus. Klein.
Rallus Virginianus. Briff.
The Little American Water-Hen. Edw.
The American Rail, or Soree. Catesby.
The Soree Gallinule. Lath. Syn.

THIS bird, which is of the bulk of the quail, is more a-kin to the land-rail than to the water-hens. It appears to be found through the whole extent of North America, as far as Hudson's Bay, though Catesby says, that he saw it only in Virginia : its plumage, he tells us, is entirely brown. He adds, that these birds grow fat in autumn, that the savages take them by speed of foot, and that they are as much prized in Virginia, as the rice-birds in Carolina, or the ortolans in Europe.

[A] Specific character of the *Rallus Carolinus :* " It is brown, " its bridle black, its breast lead-coloured, its bill bright yellow, " its feet greenish."

The JAMAICA RAIL.

Le Rale Bidi-Bidi. *Buff.*

FIFTH SPECIES.

Rallus Jamaicenfis. Gmel. and Briff.
The leaft Water-hen. Edw. and Brown.

BIDY-BIDY is the cry of this Jamaican Rail:
it fcarcely furpaffes a petty-chaps. Its head
is entirely black; the upper fide of the neck,
the back, the belly, the tail, and the wings, are
brown, variegated with whitifh crofs rays on the
back, the rump, and the belly; the feathers of
the wing and thofe of the tail are fprinkled with
white drops; the fore fide of the neck and the
ftomach are blueifh cinereous.

The LITTLE CAYENNE RAIL.

SIXTH SPECIES.

Rallus Minutus. Gmel.
The Little Rail. Lath.

THIS pretty little bird exceeds not the petty-
chaps : the fore fide of the neck and breaſt
are white, lightly tinged with fulvous and yellowiſh ; the flanks and the tail are ſtriped tranſ-
verſely with white and black ; the ground of
the feathers on the upper ſurface is black, variegated on the back with white ſpots and lines,
and fringed with ruſty colour. It is the leaſt
of the genus.

———————

THE Rails ſeem to be ſtill more diffuſed than
varied : and nature has produced or tranſported
them over the moſt diſtant lands. Captain Cook
found them at the Straits of Magellan ; in
different iſlands of the ſouthern hemiſphere, at
Anamoka, at Tanna, and at the iſle of Norfolk.
In the Society Iſlands there are two ſpecies of
Rails ; a little black ſpotted one, *(pooa-née)* and
a little

a little red-eyed one *(mai-ho)*.   It appears that
the two *acolins* of Fernandez, which he denominates *water-quails* *, are of a ſpecies of Rails
peculiar to the great lake of Mexico.   The *co-
lins*, which might be confounded with theſe, are
a kind of partridges.

* *Hiſt. Avi. Nov. Hiſp.* cap. x. p. 16.   " Acolin or water-
" quail. As large as a ſtare . . . the under ſide of its body bright
" white, its ſides ſpotted with fulvous; its upper ſide fulvous, di-
" vided by blackiſh ſpots and bright white lines, encompaſſing four
" quills."

THE CAURAL SNIPE.

The C A U R A L E.

Ardea-Helias *. Gmel. and Pallas.
Scolopax-Helias. Lath. Ind.
The Caurale Snipe. Lath. Syn.

I F we attended only to the fhape of the bill and feet, we fhould reckon this bird a rail; but its tail is much longer, and we have therefore adopted a compound name, expreffive of this character, *Caurâle* or *Queue Râle* (tail-rail). Its plumage is rich, though the colours are dark: to form an idea of it, we may compare it to the wings of thofe fine fhining flies, in which black, brown, rufous, fulvous, and light gray, inter-mingled in zones and zig-zags, compofe a foft enchanting mixture. Such particularly is the plumage of the wings and tail; the head is hooded with black, and there are long white lines above and below the eye; the bill is ex-actly that of the rail, except that it is rather longer; and the head, the neck, and the body are alfo longer than in the rail; the tail is five inches, and projects two beyond the wings; the foot is thick, twenty-fix lines high, and the

* In Cayenne it is called *petit Paon des rofes* (little peacock of the rofes.)

naked

naked part of the leg ten : the rudiment of a membrane is broader and more apparent than in the rail. The total length, from the point of the bill, which is twenty-feven lines, to that of the tail, is fifteen inches.

This bird has not hitherto been defcribed, and was but lately difcovered. It is found, though rarely, in the interior parts of Guiana, where it inhabits the fides of rivers : it lives folitary, and makes a flow plaintive whiftle, which is imitated to decoy it.

THE COMMON GALLINULE.

The WATER HEN.

LA POULE D'EAU. *Buff.*

Fulica-Chloropus. Linn. and Gmel.
Gallinula Chloropus Major. Aldrov. Johnſt. Sibb. &c.
Gallinula-Chloropus. Lath. Ind.
Gallinula. Briſſ.
The Common Water Hen, or Moor Hen. Will. Penn. &c.
The Common Gallinule. Lath. Syn. *

NATURE paſſes by gradations from the ſhape of the rail to that of the Water Hen, whoſe body is alſo compreſſed at the ſides, its bill of a ſimilar form but ſhorter, and in this reſpect liker that of the gallinaceous tribe : its head too is bare, and covered with a thick membrane ; a character of which veſtiges may be found in certain ſpecies of rails †. It flies likewiſe with its feet hanging down : its toes are extended in the ſame manner as thoſe of the rails, but are furniſhed their whole length with a membranous edging ; and this is the intermediate ſhade between the birds with cloven feet and thoſe with webbed feet.

The habits of the Water Hen correſpond to

---

* In German *Rohtblaſchen :* in Poliſh *Kokoſka.*
† Willughby.

its

its conformation: it is oftener in the water than the rail, though it does not swim much, but only crosses from one side to another. It lurks the greatest part of the day among the reeds, or under the roots of alders, willows, and oziers, and leaves not its retreat until evening: it frequents less the marshes and bogs than the rivers and pools. Its nest is placed close to the brink of the water, and constructed with a large heap of broken reeds and rushes interwoven. The mother quits her nest every evening, having previously covered the eggs with herbs and rushes. The young run as soon as they are hatched, like those of the rail, and in the same way are led by their dam to the water: and, no doubt, it is for this reason that the parents, consulting future convenience, always build their nest so near the surface. So well is the little family conducted and concealed, that it is difficult to rob it during the very short term of its education *: for the young ones are soon able to shift for themselves, and leave their prolific mother sufficient time to rear a second brood. It is even averred that they often have three hatches a year †.

The Water Hens quit the cold hilly parts in

---

* "The Water Hens conceal their young so well, that I have "never seen them, though I have fowled much in marshes at all "seasons." *Note of M. Hebert.*

† Willughby.

October,

October *, and fpend the whole winter in our temperate provinces, where they are found near fountains and uncongealed frefh waters †. Thus it can fcarcely be reckoned a bird of paffage, fince it remains the whole year in feveral countries, and only flits between the mountains and the plains.

But though the Water Hen is not migratory, and is every where fcarce, it has been planted by nature in moft of the known regions, even the remoteft. Captain Cook found it in the ifle of Norfolk ‡, and in New Zealand §: Adanfon, in an inlet at Senegal: Gmelin, in the plain of Mangafea in Siberia, near the Jenifea, where the fpecies is very numerous. Nor are thefe birds lefs common in the Antilles, at Guadeloupe ‖, at Jamaica ¶, and in the ifle of *Aves,* though it contains no frefh water: many are

* Obfervations made in the Lorraine Vofges, by M. Lottinger.
† Obfervations made in Brie, by M. Hebert.
‡ Second Voyage, *vol.* ii.
§ " The water or wood hens, though numerous enough here, " are fo fcarce in other parts, that I never faw but one. The rea- " fon may be, that, as they cannot fly, they inhabit the fkirts of " the woods, and feed on the fea-beach; and are fo very tame or " foolifh as to ftand and ftare at us till we knocked them down with " a ftick. The natives may have, in a manner, wholly deftroyed " them. They are a fort of rail, about the fize and a good deal " like a common dunghill hen; moft of them are of a dirty or dark " brown colour, and eat very well in a pye or fricaffee."—*Cook's fecond Voyage,* vol. i. p. 97.
‖ Dutertre, *tom.* ii. *p.* 277.
¶ Sloane, Browne.

found

found alſo in Canada *. And in Europe they inhabit England, Scotland †, Pruſſia ‡, Switzerland, Germany, and moſt of the provinces of France. It is true that we are not certain whether all thoſe mentioned by travellers are of the ſame ſpecies with ours. Le Page du Pratz expreſsly ſays, that the Water Hen of Louiſiana is the ſame with that of France §; and it appears that the one deſcribed by Father Feuillée at the iſland St. Thomas is nothing different ‖. We may diſcriminate, however, three ſpecies or varieties of Water Hens, which never, we are aſſured, contract affinity with each other, though they haunt the ſame pools. Thoſe found in Europe are diſtinguiſhed by their ſize, and the middle ones are the moſt common : they are about the bulk of a pullet ſix months old ; the length from the bill to the tail is a foot, and from the bill to the nails fourteen or fifteen inches ; the bill is yellow at the point, and red at the baſe ; the membranous ſpace on the front is alſo red, and ſo is the lower part of the thigh above the knee ; the feet are greeniſh ; all the plumage is of a dull iron gray, clouded with white under the body, and greeniſh brown gray above ; a white line borders the wing ; the tail,

* Hiſt. Gen. des Voy. *tom.* xv. *p.* 227.
† Rzaczyeſki.
‡ Geſner.
§ Hiſtoire de la Louiſiane, *tom.* ii. *p.* 117.
‖ Journal d Obſervations *(edit.* 1725.) *p.* 393.

when

when raifed, fhows white on the lateral feathers of the inferior coverts ;—the plumage is thick, compact, and clothed with down. In the female, which is rather fmaller, the colours are lighter, the white waves on the belly are more diftinct, and the throat is white: the fpace on the forehead is, in young fubjects, covered with a down more like hair than feathers. A young Water Hen, which we opened, had in its ftomach portions of fmall fifh and aquatic plants mixed with gravel: the gizzard was very thick and mufcular, like that of the domeftic hen: the bone of the *fternum* appeared to us much fmaller than ufual in birds; and if this difference was not owing to the age, it would partly confirm the affertion of Belon, that the *fternum* and *ifchium* are of a different fhape in the Water Hen from the fame bones in other birds.

[A] Specific character of the Water Hen, *Fulica Chloropus*: " Its front is fulvous, its bracelets * red, its body blackifh." Linnæus fays, that it has two hatches annually, and lays feven eggs about two inches long, of an ochry white colour, with a few fcarlet fpots.

* *i. e.* the coloured rings above the knees.

The LITTLE WATER-HEN.

LA POULETTE D'EAU. *Buff.*

Fulica Fufca. Linn. and Gmel.
Gallinula Fufca. Lath. Ind.
Gallinula Minor. Briff.
The Brown Gallinule. Lath. Syn.

THOUGH Belon has applied to this bird the
diminutive *poulette*, it is not much fmaller
than the preceding. Its colours are nearly the
fame; only that naturalift remarks, that it has a
blueifh tint on the breaft, and that its eye-lid is
white; he adds, that its flefh is very tender, and
that its bones are thin and brittle. We had one
of thefe birds which lived only from the 22d of
November to the 10th of December: water indeed was its only fupport: it was fhut in a narrow corner, and taken out every day by two
panes which opened in the door; at earlieft dawn
it repeatedly darted at thefe glaffes: during the
reft of its time it concealed itfelf as much as
poffible, holding down its head: if taken in the
hand, it pecked with its bill, but feebly. In this
rigorous confinement it was never heard to utter
a fingle cry. Thefe birds are in general very taciturn; they are even faid to be dumb, but when
at liberty, they have a flender call—*bri, bri, bri.*

[A] Specific character of the Brown Gallinule, *Fulica Fufca:*
" Its front is yellowifh, its bracelets of the fame colour, its body
" dufkifh."

The PORZANA, or the GREAT WATER-HEN.

Fulica Fufca, var. Gmel.
Gallinula Major. Briff.
Rallus Italorum. Johnft. Ray. Will. &c.

THIS bird is very common in Italy, in the neighbourhood of Bologna, where the fowlers call it *Porzana*. Its length from the bill to the tail is near a foot and a half; the upper fide of the bill is yellowifh, and the point blackifh; the neck and head are alfo blackifh; the upper furface is chefnut-brown; the reft of the plumage is the fame with that of the common water hen, with which we are affured it is fometimes found in our pools: the colours of the female are paler than thofe of the male.

The G R I N E T T A.

Fulica Nævia. Gmel.
Gallinula Fulica.   Lath. Ind.
Porphyrio Nævius. Briff.
Poliopus.   Aldrov. Gefner, and Ray.
The Small Water Hen.   Albin.
The Grinetta Gallinule.   Lath. Syn.

A CCORDING to Willughby, this bird is
ſmaller than the rail, and its bill is very
ſhort.   If we may judge from its different
names, it muſt be well known in the Milaneſe *.
It is found alſo in Germany, according to Geſner: that naturaliſt ſays nothing more than that
its feet are gray, its bill partly red, partly black,
the upper ſurface rufous brown, and the under
ſide of the body white.

* At Milan, ſays Aldrovandus, it is called *Grugnetta*; at Mantua *Porzana*; at Bologna *Porcellana*; and elſewhere *Guardella Co-
lumba:* at Florence it is denominated *Tordo Gelſemino*, according to
Willughby.

The S M I R R I N G.

Fulica Flavipes. Gmel.
Gallinula Flavipes. Lath. Ind.
Porphyrio Rufus. Briff.
Gallinula Ochropus Major. Ray and Will.
The Yellow-legged Gallinule. Lath, Syn,

T H E name *Smirring*, which Gefner fuppofes to have been given in imitation of the call, is in Germany the appellation of a bird which appears akin to the water hens. Rzacynfki, mentioning it as a native of Poland, fays, that it haunts the rivers, and neftles among the bufhes which grow on their banks : he adds, that the fwiftnefs with which it runs made him fome-times term it *trochilus*. In another place, he defcribes it like Gefner : " The ground of its " plumage is rufous ; the fmall feathers of the " wing are brick-colour ; the great quills of the " wing are black ; fpots of the fame are fprink- " led on the neck, the back, the wings, and the " tail ; the feet and the bafe of the bill are yel- " lowifh."

The G L O U T.

Fulica Fiftulans. Gmel.
Gallinula Fiftulans. Lath. Ind.
Porphyrio Fufcus. Briff.
The Piping Gallinule. Lath. Syn.

T H I S is a water hen, according to Gefner.
He fays that it has a fhrill high voice like
the tone of a fife : it is brown, with a little
white on the point of the wings : it is white
round the eyes, at the neck, on the breaft and
the belly ; its feet are greenifh, and its bill is
black.

FOREIGN BIRDS,

WHICH ARE RELATED TO THE WATER-HEN.

The GREAT WATER-HEN
of CAYENNE.

Fulica Cayenenſis. Gmel.
The Cayenne Gallinule. Lath.

THIS bird approaches the heron by the
length of its neck, and removes from the
water-hen by the length of its bill. It is the largeſt
of the genus, being eighteen inches long: the
neck and the head, the tail, the lower belly,
and the thighs, are brown gray; the upper ſurface is dull olive; the ſtomach and the quills of
the wings are rufous inclined to reddiſh. It is
very common in the ſwamps of Guiana, and is
ſeen even in the ditches of the town of Cayenne: it lives on ſmall fiſh and aquatic inſects:
when young its plumage is entirely gray, which
becomes reddiſh after moulting.

The MITTEK.

THE accounts of Greenland mention this
bird as a water hen, but it may be some
species of diver or grebe. In the male, the
back and neck are white; the belly black, and
the head verging on violet. In the female, the
plumage is yellow, mixed and edged with black,
so as to appear gray at a diftance. Thefe birds
are very numerous in Greenland, efpecially in
winter: they are feen flying in the morning from
the bays to the iflets, where they fubfift on
fhell-fifh; and in the evening they return to
their retreats, where they pafs the night. They
follow the windings of the coaft in their flight,
and the finuofities of the ftraits between the
iflets. They feldom fly over land, unlefs the
force of the wind, particularly when it blows
from the north, confines their excurfions. The
fportfmen feize this opportunity to fire at them
from fome promontory; thofe that are killed are
picked up by a canoe, for fuch as are wounded
go to the bottom, and never more appear *.

* Hiftoire Generale des Voyages, *tom.* xix. p. 44.

The KINGALIK.

THIS is alfo a native of Greenland, and faid to be a water-hen. It is larger than the duck, and remarkable for the indented protuberance which grows on the bill between the noftrils, and which is of an orange yellow. The male is entirely black, except that its wings are white, and its back mottled with white: the female is brown.

———

THESE are all the foreign fpecies which we can refer to the water-hens; for thofe termed *clucking hens* by Dampier are, according to his own account, akin to the herons *. Alfo the beautiful water-hen of Buenos-Ayres, defcribed by Father Feuillée, is really of a different kind, fince *its feet are like a duck's.*

* " The clucking hens refemble much the crab-eaters, but their
" legs are not quite fo long; they keep conftantly in the wet marfhy
" places, though their foot is formed like that of land birds: they
" ufually cluck like a hen with her chickens, for which reafon the
" Englifh call them *clucking hens.* There are many of them in the
" bay of Campeachy, and in other parts of the Weft Indies ... The
" crab-eaters, the clucking hens, and the goldens, with regard to
" figure and colour, refemble our Englifh herons, but are fmaller."
Dampier's Voyage round the World.

§   Laftly,

Laſtly, the *Barbary water-hen*, with ſpotted wings, of Dr. Shaw, *which is leſs than a plover*, appears to us more related to the rails, than to the water-hens *.

* *Rallus Barbaricus.* Gmel.
The Barbary Rail. Lath.

THE CHESNUT JACANA.

The JACANA.

FIRST SPECIES.

Parra-Jacana. Linn. and Gmel.
Jacana Armata Fusca. Briss.
Anser Chilensis. Charleton.
The Spur-winged Water-Hen. Edw.
The Chesnut Jacana. Lath.

" **T**HE Jacana of the Brazilians," says Marcgrave, " must be ranged with the wa-
" ter-hens, which it resembles in its instincts,
" in its habits, in the round shape of its body,
" in the form of its bill, and in the smallness of
" its head." Yet it appears to us to differ essentially from these birds by singular and even
peculiar characters: it has spurs on the shoulders, and shreds of membranes on the fore side
of the head; its toes and nails are extremely
long; the hind-toe is as long as the fore-toe;
all the nails are straight, round, and drawn out
like needles; and from this circumstance probably it received at St. Domingo the appellation
of *Surgeon*. The species is common in all the
marshes of Brazil; and we are assured that it
occurs also in Guiana and St. Domingo. We
may presume that it is likewise found in all the

tropical parts of America, both on the continent and in the iflands, as far as New Spain;
though Fernandez feems to fpeak of it only
from report, fince he makes it come from the
north, whereas it is really a native of the
fouth.

We know four or five Jacanas, which are of
the fame bulk, and differ only in colour. The
firft fpecies given by Fernandez is the fourth of
Marcgrave. The head, the neck, and the fore-
fide of the body of this bird, are black tinged
with violet; the great quills of the wing are
greenifh; the reft of the upper furface is fine
chefnut, with a purplifh or ferruginous caft:
each wing is armed with a pointed fpur inferted
in the fhoulder, exactly like the fpines of the
crifped ray-fifh; a membrane, taking its origin
at the root of the bill, fpreads on the front, and
divides into three portions, leaving alfo a barbel
on each fide; the bill is ftraight, inflated fomewhat at the point, and of a fine yellow jonquil,
like the fpurs; the tail is very fhort, and this
character, as well as the form of the bill, the
tail, the toes, and the height of the legs, of which
the half is covered with feathers, belongs equally
to all the fpecies of Jacanas. Marcgrave feems
to exaggerate, when he compares their bulk to
that of a pigeon; for their body is not larger
than the quail, only their legs are much
taller; their neck is alfo longer, and their head
                                    fmaller:

fmaller: they are always very lean; yet it is faid their flefh is palatable.

The firft fpecies of Jacana is pretty common at St. Domingo, whence it was fent us by M. Lefebure Defhayes, under the appellation of *Chevalier mordoré armé* *.

" Thefe birds," he fays, " go commonly in " pairs, and when feparated by fome accident " they call each other: they are very wild, and " the fportfman cannot approach them but by " wiles, covering himfelf with leaves, or running " behind the bufhes or the reeds. They are " feen regularly in St. Domingo during the " rainy months of May and November, or " fhortly after: yet a few are feen at other " times, which would fhew that the places of " their habitual abode are not very remote. " But they are never found except in marfhes, " or at the fides of pools and brooks.

" The flight of thefe birds is not lofty, but " pretty rapid: in rifing they vent a fhrill, " fqueaking cry, which is heard far, and feems " to bear fome refemblance to that of the white " owl. The poultry are alarmed, taking it for " the fcream of a bird of prey, though the " Jacana is very remote from that tribe. Na- " ture we might fuppofe has armed it for war, " yet we know not any foe which it combats."

This analogy to the armed lapwings, which are quarrelfome and noify birds, and have a

* *i. e.* The armed ferruginous horfeman.

fimilar

fimilar form of bill, feems to have induced fome
naturalifts to clafs them together *.　But they
differ in the fhape of their body and of their
head, and in fo far refemble the water-hen, from
which, however, they are diftinguifhed by the
peculiar conformation of their feet.　The Jacanas
may therefore be reckoned a feparate genus, appropriated to the new continent.　Their abode,
and their ftruĉture, fufficiently fhew that they
live and feed after the manner of the other fhore-
birds.　And though Fernandez fays that they
frequent only the falt bafons near fea-mark, it
appears from the above quotation that they occur in the interior parts of the country, on the
verge of frefh waters.

* Adanfon.　See the Supplement of the Encyclopedie, article
Aguapeca.

[A] Specific charaĉter of the Chefnut Jacana, *Parra-Jacana:*
Its hind nails very long, and its feet greenifh."

The BLACK JACANA.

SECOND SPECIES.

Parra Nigra. Gmel.
Jacana Armata Nigra. Briff.

ALL the head, the neck, the back, and the
tail, are black; the top of the wings and
their points are brown, the reft is green, and
the under fide of the body is brown; the fpurs of
the wing are yellow, and fo is the bill, from the
root of which a reddifh membrane rifes over the
front. Marcgrave gives this fpecies for a native
of Brazil.

The GREEN JACANA.

THIRD SPECIES.

Parra Viridis. Gmel.
Jacana. Pifon. Johnft. and Briff.

MARCGRAVE extols the beauty of this
bird, which he reckons the firft fpecies
of its genus: its back, its wings, and its tail, are
tinged with green on a black ground, and the
neck gliftens like that of a pigeon: the head is
invefted with a membrane of Turkey blue: the
bill and the nails are vermilion in their firft half,
and yellow at the point. The analogy leads us
to fuppofe that this fpecies is armed as well as
the reft, though Marcgrave does not exprefs it.

The JACANA-PECA.

FOURTH SPECIES.

Parra Brasiliensis. Gmel.
Jacana Armata. Briss.
Aguapecaca. Marcg. Johnst. Ray, Will. &c.
The Brasilian Jacana. Lath.

THE Brazilians call this bird *Agua-pecaca:*
we term it *Jacana-Peca,* to suggest both
its genus and its species. It differs little from
the preceding: " Its colours," says Marcgrave,
" are more dilute, and its wings browner ; each
" wing is armed with a spur, which serves as a
" weapon of defence; but its head is not co-
" vered with membrane." The name *Porphyrion,* which Barrere has given to this bird, seems
intended to denote its red feet. The same author says that it is common in Guiana, where
the Indians call it *Kapoua ;* and we apprehend
that the following note of M. De la Borde refers
to it. " The little species of water-hen, or *sur-
" geon,* with armed wings, is very common in
" Guiana, where it inhabits the pools of fresh
" water and the meres: it is usually seen in
" pairs, though sometimes twenty or thirty
" flock together. There are always some in
" summer

" fummer in the ditches round the town of
" Cayenne; and in the rainy feafon they appear
" even in the open parts of the new town:
" they lurk among the rufhes, and live on fifh
" and water infects." It would feem that in
Guiana, as well as in Brazil, there are feveral
fpecies or varieties of thefe birds, which are
known under different names. Aublet informs
us, that the furgeon-bird is pretty common in
Guiana, in the meres, the bafons, and the
plafhes of the favannas; that it fits on the
broad leaves of the water-lily; and that the
natives give it the appellation of *kinkin*, expreffive of its fhrill note.

The VARIEGATED JACANA.

FIFTH SPECIES.

Parra Variabilis. Gmel.
Rallus digitis triuncialibus. Klein.
Fulica Spinofa. Linn.
Jacana Armata Varia. Briff.
The Variable Jacana. Lath.

THIS Jacana has the fame predominant colours with the others, but more varied: it is greenifh, black, and purple chefnut: on each fide of the head there is a white bar, which paffes above the eyes: the fore fide of the neck is white, and alfo the whole of the under fide of the body: the front is covered with an orange-red membrane, and it has fpurs on the wing. This bird was fent to us from Brazil: Edwards reprefents one brought from Cartha-gena; which confirms our remark, that the Ja-canas are common to different parts of America, fituated between the tropics.

The SULTANA HEN, or PORPHYRION.*

Fulica-Porphyrio. Linn. and Gmel.
Gallinula-Porphyrio. Lath. Ind.
Porphyrio. Gefner. Aldrov. Johnft. Will. &c.
The Purple Water-hen. Edw. and Alb.
The Purple Gallinule. Lath. Syn.

T H E moderns have given the name of *Sultana Hen* to a bird famous among the ancients, under the name of *Porphyrion.* We have frequently had occafion to remark the juftnefs of the denominations beftowed by the Greeks, which generally allude to the diftinctive characters, and are therefore fuperior to the terms haftily adopted in our languages from fuperficial or inaccurate views. The prefent is an inftance. As this bird feemed to bear fome refemblance to the gallinaceous tribe, it got the name of *hen*; but as, at the fame time, it differed widely, and excelled by its beauty and port, it received the epithet of *Sultana.* But the term *Porphyrion,* indicating the red or purple tint of its bill and feet, was more juft and characteriftic: and fhould we not rebuild the fine ruins

* In Greek Πορφυριον, on account of its purple bill and feet. The Romans adopted this name.

of

THE PURPLE GALLINULE.

of learned antiquity, and reſtore to nature thoſe brilliant images and thoſe faithful portraits from the delicate pencil of the Greeks, ever awake to her beauties and her animation?

Let us therefore give the hiſtory of the Porphyrion, before we ſpeak of the Sultana Hen. Ariſtotle, in Athenæus *, deſcribes the Porphyrion to be a bird with long legs and pinnated feet, the plumage blue, the bill purple, and firmly fixed to the front, and its bulk equal to that of a domeſtic cock. According to the reading of Athenæus, Ariſtotle ſubjoined that it had five toes; which would have been erroneous, though ſome other ancient authors have alledged it. But among the moderns, Iſidorus has fallen into a much greater error, which has been copied by Albertus, who ſays that one of its feet is webbed and calculated for ſwimming, and that the other is fitted for running like the land birds: which is equally falſe and abſurd, and muſt mean nothing more than that the Porphyrion is a ſhore bird, and lives on the confines of the land and water. It appears, indeed, to be amphibious; for, in the domeſtic ſtate, it eats fruits, fleſh, and fiſh: its ſtomach has the ſame ſtructure with that of thoſe birds which live equally on animal and on vegetable food †.

* *Deipnos.* 9.
† Memoires de l'Academie des Sciences, depuis 1666 juſqu'en 1669, *tom.* iii. *partie* 3.

It

It is therefore eafily reared: it charms by its
noble port, its fine fhapes, its brilliant plumage,
enriched with intermingled tints of purple and
beryl: its difpofition is mild and peaceable: it
conforts with its domeftic companions, though
of different fpecies, and felects fome favourite
among them *.

It is alfo a pulverulent bird, like the cock:
yet it employs its feet, like a hand, to carry
food to its bill †. This habit feems to refult
from its proportions, the neck being fhort, and
the legs very tall; fo that it is fatiguing to
ftoop to the ground.—The ancients had made
moft of thefe remarks on the Porphyrion, and
it is one of the birds which they have defcribed
the beft.

Both the Greeks and Romans, notwithftanding their voracious luxury, abftained from eating
the Porphyrion. They brought it from Lybia ‡,
from Comagene, and from the Balearic iflands §,
to be fed ‖, and to be placed in their palaces and
temples, where it was left at liberty as a gueft ¶,

* See in Ælian the ftory of a Porphyrion which died of grief,
after having loft the cock its companion.

† Pliny, *lib.* x. 46.

‡ Alexander the Myndian, in Athenæus, reckons the Porphyrion
in the number of Lybian birds, and relates that it was facred to the
gods in that country. According to Diodorus Siculus, Porphyrions were brought from the heart of Syria, with other kinds of
birds diftinguifhed by their rich colours.

§ Pliny, *lib.* x. 46 and 49.

‖ Belon.

¶ Ælian, *lib.* iii. 41.

whofe

whofe noble afpect, whofe gentle difpofition, and whofe elegant plumage, merited fuch honours.

Now if we compare this Porphyrion of the ancients with our Sultana Hen, figured in N° 810, *Planches Enluminées*, it appears that this bird, which was brought to us from Madagafcar under the name of *taleve* *, is exactly the fame. The academicians, who have defcribed a fimilar one, recognized alfo the Porphyrion in the Sultana Hen. It is about two feet long from the bill to the claws: the toes are extremely long, and completely parted, without the leaft veftige of membrane: they are difpofed as ufual, three before and one behind; and Gefner was miftaken when he reprefented them as placed two and two: the neck is very fhort in proportion to the length of the legs, which are featherlefs: the feet are very long; the tail is very fhort; the bill is fhaped like a flat cone at the fides, and is pretty fhort: the laft property which characterifes this bird is, that its front is bald, like that of the coot's, and covered with a plate, which, extending to the top of the head, fpreads into an oval, and feems to be formed by the production of the horny fubftance of the bill. This is

---

* The *taleva* is a river bird of the bulk of a hen, which has many violet feathers; and its front, its bill, and its feet red. Flacourt fpeaks of it with admiration. *Hift. Gen. des Voyages*, t. viii. p. 606.—The French navigators call this bird *blue hen*. "The blue hens of "Madagafcar have bred on the ifle of France." *Remarks made in* 1773, *by the Vifcount de Querboënt.*

what

what Ariftotle expreffes in Athenæus, by faying
that the Porphyrion has its bill ftrongly attached
to its head. The academicians found two pretty
large *cæca*, which expanded into facs; and the
inflation of the lower part of the *æfophagus* feems
to fupply the place of a craw, which, Pliny fays,
is wanting in this bird.

This Sultana Hen, defcribed by the Academicians, is the firft bird of the kind that has
been feen by the moderns. Gefner fpeaks only
from report, and from a drawing of it: Willughby fays, that no naturalift has feen the Porphyrion. We owe to the Marquis de Nefle the
pleafure of having feen it alive; and we exprefs
our moft refpectful thanks for what we regard
as a debt of Natural Hiftory, which every day
is enriched by his enlightened and generous
tafte: he has put it in our power to verify in a
great meafure, on his Sultana Hen, what the ancients have faid of their Porphyrion. This bird
is very gentle and innocent, and at the fame
time timorous, fugitive, fond of folitude and
retirement, concealing itfelf as much as poffible when it eats. It cries from fear when one
approaches, at firft with a faint found, which
afterwards grows fhriller and louder, and ends
with two or three dull and hollow claps: while
in a cheerful mood, it vents fofter and calmer
accents. It feems to prefer fruits and roots,
particularly thofe of the fuccories, to every other
fort of food, though it can alfo live on feeds. If

§   offered

offered a fifh, it eagerly feizes it, and devours it greedily. Often it repeatedly foaks its provifions in water. How fmall foever its morfel may be, it conftantly clenches it with its long toes, bending the hind one over the reft, and holding its foot half raifed; it then eats by crumbs.

Scarce any bird has more beautiful colours; the blue of its plumage is foft and gloffy, embellifhed with brilliant reflections; its long feet, and the plate from the top of its head to the root of its bill, are of a fine red, and a tuft of white feathers under the tail heightens the luftre of its charming garb. Except that it is rather fmaller, the female differs not from the male, which exceeds the partridge, but is inferior to a common hen. The Marquis de Nefle brought this pair from Sicily, where, according to the note which he obligingly communicated to us, they are known under the name of *Gallofagiani*: they are found on the lake Lentini, above Catana, and are fold for a moderate price in that city, as well as in Syracufe and the adjacent towns. They appear alive in the public places, and plant themfelves befide the fellers of vegetables and fruits to pick up the refufe: and this beautiful bird, which the Romans lodged in their temples, now experiences the decline of Italy. That fact fhows that the Sultana Hens have been naturalized in Sicily from a few pairs of thefe Porphyrions introduced from

Africa;

Africa; and in all probability this fine fpecies has been. propagated, in like manner, in fome other countries; for we fee from a paffage of Gefner, that this naturalift was convinced that thefe birds are found in Spain, and even in the fouth of France.

This bird is one of thofe which are by nature moft difpofed to domeftication, and to multiply them would be both agreeable and ufeful. The pair kept in the voleries of the Marquis de Nefle, neftled laft fpring (1778); both male and female laboured in conftructing the neft; they placed it at fome height from the ground, on a projection of the wall, with a heap of fticks and ftraws: the eggs were fix in number, white, with a rough fhell, exactly round, and about the fize of a demi-billiard. The female was not affiduous in covering; and a common hen was fubftituted, but without fuccefs. We may furely expect another hatch to be more profperous if carefully attended by the mother herfelf; for this purpofe thefe birds ought to enjoy the calm and retreat which they feek, and efpecially in the feafon of love.

[A] Specific character of the purple Gallinule, *Fulica Porphyrio* : " Its front is red, many bracelets; its body green, below violet."

B I R D S,

WHICH ARE RELATED TO THE SULTANA HEN.

SINCE the primary ftock of the Sultana Hen inhabits the fouthern regions of our continent, it is not probable that the climates of the north produce the fecondary fpecies. We muft therefore reject the fourth, fifth, fixth, feventh, and eighth fpecies of Briffon, which he prefumes to have the frontal plate, though Gefner, from whom he borrowed the defcriptions, gives no indication of this plate, either in his text or by his figures. The fecond of thefe appears to be a rail, and accordingly we have ranged it in that genus : the four others are water hens, as the original author himfelf fays. With regard to the ninth fpecies of Briffon, which he calls the *Sultana Hen of Hudfon's Bay*, it ought to be excluded, both on account of the climate, and becaufe Edwards gives it as a coot, remarking at the fame time that it is more akin to the rail. Notwithftanding thefe retrenchments, there ftill remain three fpecies in the ancient continent, which feem to form the intermediate ſhade between the Sultana Hen *, the coots, and the water hens.   There are alſo three ſpecies in America, which appear the repreſentatives, in the new world, of the Sultana Hen and its ſubordinate ſpecies.

* Mr. Forſter found at Middleburg, one of the Society Iſles, *coots with a blue plumage,* which ſeem to be Sultana Hens.

The GREEN SULTANA HEN.

FIRST SPECIES.

Fulica Viridis. Gmel.
Gallinula Viridis. Lath. Ind.
Porphyrio Viridis. Briſſ.
The Green Gallinule. Lath. Syn.

THIS bird is much ſmaller than the Sultana, and exceeds not a rail.   All the upper ſide of the body is dull green, but gloſſy; and all the under ſide of the body white, from the cheeks and the throat to the tail: the bill and frontal plate are yellowiſh-green.   It is found in the Eaſt Indies.

The BROWN SULTANA HEN.

SECOND SPECIES.

Rallus Phœnicurus, var. 1. Gmel.
Gallinula Phœnicura, var. 1. Lath.

THIS bird comes from China : it is fifteen or sixteen inches long. It has none of the rich tints that seem peculiar to this genus of birds, and perhaps the specimen is a female. All the upper side of the body is brown or blackish cinereous; the belly is rufous; the fore side of the body, of the neck, and of the throat, and the circle about the eyes, are white; the frontal plate is small, and the bill varies somewhat from the conical shape which obtains in the true Sultana ; it is longer, and resembles more the bill of the water hens.

The ANGOLI.

THIRD SPECIES.

Fulica Maderaspatana. Gmel.
Porphyrio Maderaspatanus. Briss.
Gallinula Maderaspatana. Lath. Ind.
Crex Indica. Ray.
The Madras Rail-hen. Id.
The Madras Gallinule. Lath. Syn.

THIS bird is commonly at Madras called *Caunangoli,* which we have shortened into *Angoli* ; the Gentoos term it *Boollu-cory.* It is

difficult

difficult to determine whether it ought to be referred to the fultanas, the water hens, or even the rails : all that we know of it is a fhort hint given by Petiver in his addition to Ray's Synopfis ; but this indication, like all the others of that fragment, is formed from drawings fent from Madras, and exprefles not the difcriminating charaċters. Briflon makes it his tenth fpecies of Sultana Hen, and by confequence prefumes that it has the frontal plate, though Petiver never mentions it : on the contrary, he fays, that its bill is flender, fharp, and longifh ; he applies the names of *crake* and *rail,* and he reprefents it as equal in bulk to a goofe. So far it refembles more the fultana : and this is all that we can fay, till we are better informed.

The LITTLE SULTANA HEN.

FOURTH SPECIES.

Fulica Martinica. Linn. and Gmel.
Porphyrio Minor. Briff.
Gallinula Martinica. Lath. Ind.
The Martinico Gallinule. Lath. Syn.

THE genus of the Sultana Hen occurs, as we have faid, in the new world ; and if the fpecies are not exaċtly the fame, they are at
leaft

leaſt analogous. The preſent is a native of Guiana: it is only ſomewhat larger than the water-rail. It reſembles our Sultana Hen ſo cloſely, that in the whole hiſtory of birds there are few examples of analogies ſo intimate between thoſe of the two continents. Its back is blueiſh green; and all the fore ſide of the body is ſoft violet-blue, which covers alſo the neck and head, aſſuming a deeper caſt. It appears to us to be the ſame with what Briſſon makes his ſecond ſpecies.

The F A V O U R I T E.

FIFTH SPECIES.

Fulica Flaviroſtris. Gmel.
Gallinula Flaviroſtris. Lath. Ind.
The Favourite Gallinule. Lath. Syn.

THIS is nearly of the ſame ſize with the preceding, and comes from the ſame country. Perhaps it is only the female of the ſame ſpecies, eſpecially as the colours are the ſame, only more dilute; the blueiſh green of the wings, and the ſides of the neck, are faint; brown ſhines through on the back and on the tail: all the fore ſide of the body is white.

The ACINTLI.

SIXTH SPECIES.

Fulica Purpurea. Gmel.
Gallinula Purpurea. Lath. Ind.
Quachilton. Fernand. Nieremb. Johnſt. &c.
The Crowing Gallinule. Lath. Syn.

THIS Mexican bird, which Briſſon refers to our ſultana hen, or porphyrion of the an-cients, differs by ſeveral characters : beſides that we can hardly ſuppoſe that a bird of ſuch laborious flight could paſs from the one conti-nent into the other; the toes and feet of the Acintli are not red, but yellow or greeniſh ; all its plumage is blackiſh purple, intermingled with ſome white feathers. Fernandez gives it the names of *quachilton* and *yacacintli*; the lat-ter of which we have adopted, and ſhortened. The denomination of *avis ſiliquaſtrini capitis* is very expreſſive, and ſhows that the flat frontal plate is like a large pod, a character which con-nects this bird with the coots and ſultana hens. Fernandez adds, that the Acintli crows like a cock during the night and at the break of day ; which might afford a ſuſpicion that it belongs not to the genus of the ſultana hen, whoſe voice bears no reſemblance to that of the cock.

8

A bird

A bird of a species nearly allied to this, if not the same, is described by Father Feuillée, under the name of *water hen* *: it has the character of the sultana, the broad flat escutcheon on the front; all its attire is blue, except a cowl of black on the head and neck. Feuillée remarks also differences of colours between the male and female, which occur not in our sultana hens, in which the female is smaller than the male, but both perfectly alike in colours.

Nature has therefore produced at great distances the species of sultana hens, but always in the southern latitudes. Forster found it in the South Sea; and the *purple water hen* which he saw at Anamooka appears to be a bird of the same family.

* " The female has its crown deep fulvous; its mantle of the " same colour; its facings white; its wings greenish, mixed with a " little fulvous; the quills sky-blue, mixed with a little green: " these birds are very lean, and have a disagreeable marshy taste." *Feuillée.*

[A] Specific character of the Crowing Gallinule, *Fulica Purpurea:* " It is purple, its bill pale, its feet greenish-yellow."

The COMMON COOT.

LA FOULQUE. *Buff.*

Fulica Atra. Linn. and Gmel.
* *Fulica.* All the Naturalists.

THE species of the Coot commences the ex-
tensive tribe of true aquatic birds. Though
its feet are not completely webbed, it lives ha-
bitually on the water, and seems even more attached to that element than any fowl, except the
diver. It is seldom seen on land, and is there
so bewildered and defenceless, that it frequently
suffers itself to be caught with the hand. It
spends the whole day on the pools, which it
prefers to the rivers; and, except in walking
from one pool to another, it never sets foot on

---

* The Greek name is conjectured, from a passage in Aristotle,
lib. ix. 35, to be Φαλαρις; the modern Greeks call it Λυφα: in
Latin *Fulica* or *Fulix,* because of its dusky colour; *fuligo,* smoke:
hence the Italian *Follega* or *Follata:* on Lago Maggiore *Pullon:* in
Catalonia *Folge, Follaga, Gallinosa de Aigua* (water hen): in Germany *Wasser-houn, Rohr-heunle* (reed hen), *Tautcherlein* (diver):
in Suabia *Blesz, Blessing:* in Lower Saxony *Zapp:* in Switzerland
Belch, Bellique, Belchisen: in Holland *Meer-Coot:* in Sweden *Blaos-
Klacka:* in Denmark *Blis-hone, Blas-and, Vard-hone:* in Poland
Lyska, Dzika, or *Kacza:* in many provinces of France *Sudelle*; and
in Picardy *Blerie.*

shore :

THE COMMON COOT.

shore: and if the interval be confiderable, it has recourfe to its wings, and rifes very high; but commonly it flies only in the night *.

The Coots, like many other water fowl, fee beft in the dufk, and the older ones never feek their food but in the night †. They lurk among the rufhes the greateft part of the day; and when difturbed in their retreat, they will bury themfelves in the mud rather than fpring. They feem to make an effort in commencing the motion fo natural to other birds; and whether on water or on land they rife with difficulty. The young Coots, lefs folitary or circumfpect, are feen at all hours of the day bouncing with fmall leaps out of the water, one fronting another. They fuffer the fowler to approach, yet eye him fteadily; and they plunge fo nimbly, the inftant they perceive the flafh, that often they elude the fhot. But in autumn, when thefe birds leave the fmall pools and affemble on the lakes, vaft quantities are caught ‡. For this purpofe, a number of fkiffs are arranged in a line extending the breadth of the lake; this little fleet is rowed

* " I never faw it fly during the day but to avoid the fowler; " but I have heard it pafs over my head at all hours of the night." Obfervation of M. Hebert.

† According to Salerne, the Coot, when other food fails (and this can feldom happen) dives, and tears up from the bottom of the water the root of a great rufh, which it gives its young to fuck.

‡ Particularly in Lorraine, on the great pools of Tiaucourt and of Indre.

forward,

forward, and drives the Coots into some inlet:
the birds, then, urged by fear and necessity, rise
at once into the air, and, endeavouring to regain
the open water, they pass over the heads of the
fowlers, and receive a general and destructive
fire. The same plan of operation is now conducted at the other end of the lake, where those
which escaped have alighted; and what is singular, neither the clamours of the sportsmen,
the report of the muskets, the spectacle of the
range of boats, nor the death of their companions, can induce these birds to betake to distant flight. They do not quit this scene of carnage till the night following; and a few linger
behind next morning.

These indolent birds have deservedly many foes:
the moor buzzard sucks their eggs, and plun-
ders their young; and to this destruction must
be imputed the fewness of their number, considering that they are very prolific. The Coot
lays eighteen or twenty eggs, which are of a
dirty white, and almost as large as a hen's; and if
the first hatch be destroyed, the mother has often
a second, of ten or twelve eggs *. She builds
in deluged spots covered with dry reeds: she
selects a tuft, on which she raises a structure
above the level of the water, and lines the cavity with little dry herbs and tops of reeds,
forming a large shapeless nest, distinguishable at

Observation of M. Baillon.

a dis-

a diftance *.   She fits twenty-two or twenty-
three days, and as foon as the young are hatched
they jump out of the neft, and never return
again.   The mother cherifhes them under her
wings, and they fleep round her beneath the
reeds : fhe leads them to the water, in which
they fwim and dive well, from the moment of
their birth.   They are covered at firft with a
fmoky black down, and look very ugly ; only
the trace is to be feen of the white plate deftined to ornament their front.   It is then that
the bird of prey affails them fo cruelly, and often
devours the dam and her brood †.   The old
Coots, which have repeatedly loft their callow
offspring, grow cautious from misfortune, and
conceal their nefts among the flags on the margin of the pools ; and keep together their young
among thefe thick coverts.   Thefe alone perpetuate the fpecies ; for fo great is the depopula-
tion of the reft, that a good obferver, who has
particularly ftudied the œconomy of the Coots ‡,
reckons that not above one-tenth efcape the
talons of the birds of prey, particularly thofe of
the moor buzzard.

* There is little probability that the Coot, as Salerne alledges,
makes two nefts, one for hatching, and another for lodging her
young. What may have given rife to this notion is, that the brood,
after they have once quitted the neft, never return to it, but fquat
with their mother among the rufhes.

† The fame Salerne pretends, that the Coot defends itfelf againft
the bird of prey, by prefenting its talons, which are, indeed, pretty
fharp ; but this feeble refiftance muft generally be of little avail.

‡ M. Baillon.

The

The Coots breed early in the spring, and eggs are found in their body as soon as the end of winter *. They reside on our pools the greatest part of the year, and in some places they are permanent settlers †. Yet in autumn they all leave the small pools, and resort to the large ones, where they assemble in a great flock: there they often remain till December; and when the snows, and especially the frosts, drive them from the high and chill tracts, they descend into the plains, which enjoy a milder temperature; and the want of water, rather than the cold, constrains them to shift their haunts. M. Hebert saw them in a very severe winter on the lake of Nantua, which is late in freezing: he saw them also in the plains of Brie, though in small numbers, in the depth of winter. But, most probably, the bulk of the species remove by degrees to the adjacent countries, which are warmer: for, as their flight is laborious and tardy, they cannot journey to any great distance; and indeed they appear again as early as February.

The Coots are spread through the whole of Europe, from Italy to Sweden: they are found, too, in Asia ‡. They occur in Greenland, if Egede rightly translates two words in the lan-

* Belon.

† As in Lower Picardy, according to the observations of M. Baillon.

‡ Lettres Edifiantes, *thirtieth collection, p.* 317.

guage

guage of the natives, by the great and little
Coot*.   In fact, the species consists of two families, which live in the same lake without ever
cohabiting, and are distinguished from each other
solely by their bulk, and not by the colour of
the frontal plate, as some pretend; for in both
that is usually white, and becomes red only in
the season of love.

This thick naked membrane, which covers
the fore side of the head like an escutcheon,
and which made the ancients give the Coot the
epithet of *bald*, seems to be a production of the
upper layer of the substance of the bill, which is
soft, and almost fleshy near the root.   The bill
is fashioned into a flat cone at the sides, and is
bluish white; when in the season of courtship,
the frontal plate assumes its vermilion tint.

All the plumage is furnished with a thick
down, covered with delicate close feathers; it is
of a leaden-black, full and deep on the head and
neck, with a white streak on the fold of the
wing: no difference indicates the sex.   The
Coot is as large as a domestic hen, and its head
and body are nearly of the same form: its toes
are half-webbed, fringed fully on both sides
with a membrane, scalloped into festoons, whose
knots correspond to the joints of the *phalanges:*
these membranes are, like the feet, of a leaden
colour: above the knee a small portion of the
naked leg is circled with red: the thighs are

* *Navia* and *Navialurfoak.*

thick

thick and flefhy. Thefe birds have a gizzard,
two large *cæca*, and a capacious gall-bladder *.
They live chiefly, as well as the water-hens, on
aquatic infects, fmall fifh, and leeches ; yet they
alfo gather feeds, and fwallow pebbles : their flefh
is black, lean, and has a flight marfhy tafte.

In the ftate of liberty, the Coot has two different cries, the one broken, the other drawling: it is the latter, no doubt, from which Aratus draws a prognoftic † ; for the former is reprefented by Pliny as boding ftorms ‡. But
captivity feems to difpirit and opprefs it fo
much, that it lofes its voice, and would feem
abfolutely mute.

* Belon.

† *Haud modicos tremulo fundens è gutture cantus.* Apud Cicer.
lib. i. *De Nat. Deor.*

‡ *Et fulica matutino clangore tempeftatem.* lib. xviii. 35.

[A] Specific character of the Common Coot, *Fulica Atra :* " Its
" front is flefh-coloured, its bracelets yellow, its body blackifh."
The Coots remain the whole year in Great Britain. They are
found alfo in North America : on the rivers in Carolina they are
called *flufterers.* The favages, near the falls of the Niagara, drefs
their fkins for pouches.

The GREATER COOT.

LA MACROULE, *ou* GRANDE FOULQUE.

Buff.

Fulica Aterrima. Linn. and Gmel.
Fulica Major. Will. Ray, and Briff.

ALL that we have faid of the common Coot *(morelle)* applies to the Greater Coot *(macroule)*; their habits and their fhapes are the fame, only the latter is rather larger than the former, and the bald fpace on the front is alfo broader. One of thefe birds, taken in March 1779, near Montbard, among the vines, whither it had been driven by a violent wind, afforded me an opportunity, for the fpace of a month, during which it was kept alive, of making the following obfervations. It refufed, at firft, all forts of dreffed food, bread, cheefe, and flefh, raw or boiled: it rejected alfo earth-worms and young frogs, whether dead or alive. It required to be crammed with gobbets of foaked bread. It was extremely fond of a tray full of water, and would repofe whole hours in it. It fought alfo to hide itfelf; though it was not wild, and fuffered itfelf to be laid hold of, only pecking, with a few ftrokes, the hand that was about to feize it, and

thefe

thefe fo feeble, either becaufe of the foftnefs of the bill, or the weaknefs of the mufcles, as hardly to make any impreffion on the fkin: it betrayed neither anger nor impatience; it made no endeavour to efcape, and fhewed no furprize or fear. But this ftupid tranquillity, this total want of vigour and courage, proceeded probably from its bewildered condition, remote from its proper element and its ufual habits. It feemed deaf and mute; any noife made clofe to its ear never moved it, or drew the leaft inclination of its head; and though it was often purfued and teazed, it never vented the fmalleft cry. We have the water-hen equally mute in captivity. The mifery of flavery is greater than is fuppofed, fince it fometimes bereaves its unhappy victims of even the power of complaining.

[A] Specific character of the Greater Coot, *Fulica Aterrima*: " Its front is white, its bracelets red, its body blackifh."

The CRESTED COOT.

LA GRANDE FOULQUE A CRÊTE. *Buff.*

Fulica Criftata. Gmel.

IN this Coot, the flefhy plate on the front is raifed and detached in two fhreds, which form a real comb. It is befides confiderably larger than the preceding fpecies, which it exactly refembles in its fhape and plumage. It was fent to us from Madagafcar. May it not be really the fame with the European, only expanded by the influence of a hotter and more active climate?

The PHALAROPES.

EDWARDS was the firft who introduced this genus of fmall birds, which, with the bulk and almoft the fhape of the fandpiper, have feet like thofe of the coot. From this analogy, Briffon terms them *Phalaropes* * ; while Edwards, refting on their more obvious appearance, is contented with the name *Tringa*. They are, indeed, little fnipes or fandpipers, on which

---

* From Φαλαρις, which is probably the Greek for the coot; and πυς, the foot.

　　　　P　　　　nature

nature has beſtowed the feet of the coot. They ſeem to belong to the northern countries: thoſe figured by Edwards came from Hudſon's Bay, and we received ſome from Siberia. But whether they migrate or ſtray, they are ſometimes ſeen in England; for Edwards mentions one which was killed in winter in Yorkſhire. He deſcribes four different birds, which may be reduced to three ſpecies.

The CINEREOUS PHA-LAROPE.

FIRST SPECIES.

Tringa Hyperborea, maſ. Linn. and Gmel.
Phalaropus Cinereus.
Tringa Fuſca. Gmel.
Phalaropus Fuſcus. Lath. Ind.
Phalaropus. Briſſ.
The Coot-footed Tringa. Edw.
The Brown Phalarope. Penn. and Lath.

IT is eight inches long from the bill to the tail, which projects not beyond the wings: its bill is ſlender, flattened horizontally, thirteen inches long, ſlightly ſwelled and bent near the point; its feet are deep fringed, like thoſe of the coot, with a membrane in feſtoons, whoſe knots correſpond alſo to the articulations of the toes:

4   the

the upper furface of the head, neck, and body,
is gray, waved gently on the back with brown
and blackifh: it has a white neck-piece, inclofed by an orange rufous line; below it, the
neck is encircled with gray, and all the under
fide of the body is white. Willughby fays,
that he was informed by Dr. Johnfon, that this
bird has the fhrill clamorous voice of the fea-
fwallows: but he did wrong to range it with
thefe fwallows, efpecially as he remarked its analogy to the coot.

The RED PHALAROPE.

SECOND SPECIES.

Tringa Fulicaria. Linn. and Gmel.
Phalaropus Hyperboreus, fem. Lath. Ind.
Phalaropus Rufefcens. Briff.
The Red Coot-footed Tringa. Edw.

THE fore fide of the neck, the breaft, and the
belly, are brick-coloured; the throat rufous brown, fpotted with blackifh; the bill is
quite ftraight, like that of the fandpiper; the
toes fringed with broad membranous feftoons:
it is rather larger than the preceding, being
equal to a kingfifher.

The PHALAROPE with INDENTED FESTOONS.

THIRD SPECIES.

Tringa Lobata. Linn. and Gmel.
Phalaropus Lobatus. Lath. Ind.
Phalaropus. Briff.
The Gray Coot-footed Tringa. Edw.
The Gray Phalarope. Penn. and Lath.

THE fcalloped feftoons, which were fmooth in the preceding, are here delicately indented on the edges; and this character fufficiently difcriminates it. Like the firft fpecies, it has its bill flattened horizontally, a little inflated near the point, and hollowed above by two grooves; its eyes are a little drawn towards the back of its head, whofe top bears a blackifh fpot, the reft being white, which is the colour of the whole of the fore fide and under fide of the body: the upper fide is flaty-gray, with tints of brown, and obfcure longitudinal fpots. It is of the fize of the jack fnipe.

The GREBE.

FIRST SPECIES.

Colymbus Urinator. Linn. and Gmel.
Colymbus. Briff.
Colymbus Major. Aldrov. Will. Ray, Johnft. &c.
The Greater Loon, or Arsefoot. Will.
The Greater Dobchick. Edw.
The Tippet Grebe. Penn. and Lath.*

THE Grebe is well known by thofe beautiful filvery white muffs, which have the
foft clofenefs of down, the elafticity of feathers, and the luftre of filk.   Its undreffed
plumage, particularly that of the breaft, is really
a fine down, very clofe and firm, and regularly
difpofed, whofe gliftening filaments lie upon
each other, and join, fo as to form a glaffy,
fhining furface, equally impenetrable by cold or
humidity.   This clothing, fo well adapted to
the rigours of feafon and of climate, was neceffary to the Grebe, which in the fevereft winters remains conftantly in the water, like the
divers; infomuch that it has often been confounded with them under the common name
colymbus †.   But the Grebes differ effentially
from the divers, which have their toes com-

---

* In Italian *Smergo, Fifolo Marino:* in German *Deucchel.*
† From Κολυμβαω, *to go into the water.*

pletely

pletely webbed, and not edged with a fcalloped
membrane, parted at each toe; not to mention
other diftinctions, which fhall be afterwards ftat-
ed. Accordingly, accurate naturalifts, appro-
priating to the divers the terms *mergus, uria,*
and *æthya,* reftrict that of *colymbus* to the great
and little Grebes.

By its ftructure, the Grebe is deftined to inhabit the waters; its legs are placed entirely
behind, and almoft funk into its belly, fo that
only the feet appear, and are like oars; they naturally throw themfelves outwards, and could
not fupport the body of the bird on the ground,
unlefs it ftood quite erect. In this pofition, the
ftriking with its wings would, inftead of raifing
it into the air, only overturn it; fince the legs
could not aid the impulfion. It requires therefore a great effort to begin its flight on land; and,
as if confcious of this imbecility, it is obferved
to avoid the fhore; and to prevent its being
driven thither, it always fwims againft the
wind *. If unfortunately a wave cafts it on the
brink, it continues ftruggling with its feet and
wings, though for the moft part in vain, to
mount into the air, and return to the water: it
may be then caught by the hand, in fpite of the
violent ftrokes it gives with its bill in defence.
But it is as nimble in the water as it is feeble on
land: it fwims, dives, dafhes through the waves,
and runs on the furface with furprifing rapidity;

* Oppian. *Exeutic. lib.* ii.

its

its motions are faid even to be never quicker
and brifker than when under water *. It purfues the fifh to a very great depth †, and is often caught in fifhermen's nets. It dives deeper
than the fcoter duck, which is taken only on
beds of fhell-fifh left bare by the ebb-tide;
while the Grebes are taken in the open fea,
often at more than twenty feet depth.

The Grebes frequent equally the fea and the
frefh waters, though naturalifls have fcarce
fpoken but of thofe which are feen on lakes,
pools, and inlets of rivers ‡. Several fpecies occur on the coafts of Brittany, Picardy, and in the
channel §. The Grebe of the lake of Geneva,
which is found alfo on that of Zuric, and on the
other lakes of Switzerland ||, and fometimes on
that of Nantua, and even on certain pools of
Burgundy and Lorraine, is the kind beft known.
It is rather larger than the coot; its length
from the bill to the rump is a foot five inches,
and from the bill to the nails a foot and nine or
ten inches : all the upper fide of the body is
deep brown but gloffy, and all the fore fide is of
a very fine filvery white. Like all the other
Grebes, it has a fmall head, a ftraight and pointed bill, and from the corners a fmall naked red
fkin extends to the eye; its wings are fhort and

* Willughby.
† Schwenckfeld.
‡ *Idem.*
§ The little and crefted Grebes, according to M. Baillon.
|| Gefner.

P 4   fomewhat

somewhat difproportioned to the body. The
bird rifes with difficulty, but after it has caught
the wind, it flies far *. Its voice is loud and
rough †. Its leg, or rather its tarfus, is wi-
dened and flattened laterally; the fcales with
which it is covered form on the hind part a
double indenting; the nails are broad and flat:
the tail is wanting in all the grebes, but they
have on the rump the tubercles, in which the
tail-quills are ufually inferted, but thefe tubercles are fmaller than in other birds, and only
bear a tuft of fmall feathers.

Thefe birds are commonly very fat: not only
they feed their young with little fifhes, they eat
fea-weed and other plants ‡, and fwallow mud §.
White feathers too are often found in their ftomach; not that they devour birds, they catch
the down which plays on the water, miftaking
it for a fmall fifh. It is moft probable that the
Grebes, like the cormorants, caft up the refidue
of digeftion; at leaft fifh-bones are found rolled
into pellets and unaltered at the bottom of the
ventricle.

The fifhers of Picardy refort to the Englifh
coaft in fearch of the Grebes' nefts, fince they
do not breed on that of France ‖; and they find
thefe in the holes of rocks, into which the birds

* Willughby.
† Gefner and Belon.
‡ Willughby.
§ Schwenckfeld.
‖ Obfervations of M. Baillon.

probably

probably fly, fince they cannot climb, and whence the young muſt throw themſelves into the ſea. But on our large pools they build with reeds and ruſhes interwoven, and the neſt is half dipped in the water, though not entirely afloat, as Linnæus aſſerts, but ſhut and attached to the reeds *. It commonly contains two eggs, ſeldom three. Againſt the month of June the young neſtlings are ſeen ſwimming with their dam †.

The genus of theſe birds conſiſts of two families differing in ſize. To the large ſort we ſhall appropriate the name *grebes*, and to the ſmall, that of *cheſnuts (caſtagneux)*: this diviſion is natural and ancient, and ſeems to be indicated by Athenæus under the terms *colymbis* and *colymbida*; ſince to the latter he conſtantly joins the epithet of *little*. There is however conſiderable variety in regard to ſize.

* Obſervations of M. Lottinger.
† *Idem.*

[A] Specific character of the Tippet Grebe, *Colymbus Urinator* : " Its head is ſmooth, its lower eye-lid yellow, a white ſpot on the " wings." The Grebes are very attentive in feeding their young, and will even carry them when tired on their back or under their wings. Their fleſh is rank, but their fat is ſuppoſed to have great virtue in rheumatic complaints. On the lake of Geneva, theſe birds appear in ſmall flocks of ten or twelve: they ſell fourteen ſhillings a-piece. The ſkin, with the beautiful plumage on the under ſide of the body, is made into muffs and tippets. This ſpecies is rare in England.

The LITTLE GREBE.

SECOND SPECIES.

Colymbus Obscurus. Gmel.
Podiceps Obscurus. Lath. Ind.
Colymbus Minor. Briss.
The Black and White Dobchick. Edw.
The White and Dusky Grebe. Penn.
The Dusky Grebe. Lath. Syn.

THIS is smaller than the preceding, which is almost their only difference. But if that be constant, they ought to be discriminated; especially since the little Grebe is known in the channel, and inhabits the sea-coast, whereas the great Grebe occurs more frequently in fresh waters.

[A] Specific character of the *Colymbus Obscurus:* " Its head " is smooth; its front, the under side of its body, and the tips of its " secondary wing-quills, white."

The CRESTED GREBE.

THIRD SPECIES.

Colymbus Criſtatus. Gmel. and Briſſ.
Podiceps Criſtatus. Lath. Ind.
Colymbus Major, Criſtatus & Cornutus. Ald. Geſn. &c.
Acitli. Hernandez.
The Greater Creſted and Horned Ducker. Alb. and Plot.
The Car Gooſe. Charleton.
The Gray or Aſh-coloured Loon. Will.
The Great Creſted Grebe. Penn. and Lath.

THE feathers on the crown of the head extend a little behind, and form a ſort of
creſt, which it raiſes or depreſſes according as
it is tranquil or diſcompoſed. It is larger than
the common grebe, being at leaſt two feet from
the bill to the nails : but it differs not in its
plumage ; all the fore ſide of the body being of
a fine ſilvery white, the upper ſide blackiſh
brown, with a little white on the wings.——
Theſe colours compoſe the general livery of the
grebes.

It appears from comparing the indications of
ornithologiſts, that the Creſted Grebe inhabits
equally ſeas and lakes, the coaſts on the Mediterranean, and thoſe waſhed by the Atlantic.
The ſpecies occurs even in North America,
and is the *Acitli,* which Hernandez ſays frequents
the lake of Mexico.

It has been remarked, that the Grebes of this
ſpecies, and probably it is the ſame with the
others,

others, acquire not till after moulting their fine
fatin white. The iris, which is always very
brilliant and reddifh, becomes inflamed, and
affumes a ruby tint, in the feafon of love. This
bird is faid to deftroy numbers of young whit-
ings and fturgeons' fry, and, when in want of
other food, to eat fhrimps *.

* Obfervations made in the Channel, by M. Baillon, of Montreuil-fur-mer.

[A] Specific character of the Crefted Grebe, *Colymbus Criftatus* : " Its head is rufous, its neck black, its fecondary wing-quills
" white." Thefe birds are frequent in the fens of Lincolnfhire,
and on the meres of Shropfhire, Chefhire, and Staffordfhire, where
they breed. Their fkins are made into tippets equally valuable
with thofe from Geneva. Mr. Latham reckons the Crefted Grebe
to be the adult, and the Tippet Grebe the young bird of the year's
hatch.

———————

The LITTLE CRESTED GREBE.

FOURTH SPECIES.

Colymbus Auritus. Gmel.
Podiceps Auritus. Lath. Ind.
The Eared Dobchick. Edw.
The Eared Grebe. Penn. and Lath.

THIS Grebe is not larger than a teal, and
differs from the preceding not only in fize,
but alfo becaufe the feathers on the crown of the
head, which compofe the creft, are parted into
two tufts, and that fpots of chefnut-brown mix
with the white on the fore fide of the neck.
With

THE CRESTED GREBE

With refpect to the identity fuppofed by Briffon
between this fpecies and " the greater afh-co-
" loured ducker" of Willughby, it is difficult
to decide; fince that naturalift and Ray form
their defcription merely from a drawing of
Brown's.

[A] Specific character of the *Colymbus Auritus*: " Its head is
" black, crefted with ferruginous ears."

The HORNED GREBE.

FIFTH SPECIES.

Colymbus Cornutus. Gmel.
Podiceps Cornutus. Lath.
The Eared or Horned Dobchick. Edw.

THIS Grebe has a black tuft, divided behind
as it were into two horns; it has alfo a fort
of mane, rufous at the root, black at the point, cut
round the neck; which gives it a very odd look,
and makes it to be regarded as a monftrous fpecies.
It is rather larger than the common grebe; its
plumage is the fame, except the mane and the
flanks, which are rufous.

This Horned Grebe feems to be extenfively
fpread : it is known in Italy, in Switzerland, in
Germany, in Poland, in Holland, and in England. As it is of a very fingular figure, it has
been every where remarked. Fernandez defcribes

scribes with accuracy one found in Mexico; and adds, that it is called the *water hare*, but does not assign the reason.

[A] Specific character of the *Colymbus Cornutus*: " Its head is " glossy-green, with a yellow bar at its eye, extended behind like " a crest."

The LITTLE HORNED GREBE.

S I X T H   S P E C I E S.

Podiceps Cornutus, var. Lath.
Colymbus Cornutus Minor. Briff.

THERE is the same difference with regard to bulk between the two horned grebes as between the two crested grebes. The little Horned Grebe has two pencils of feathers, which growing out from behind the eyes, form its horns of an orange rufous: this is also the colour of the fore side of the neck and of the flanks; the top of the neck is clothed with puffed feathers, not broken, however, or intersected by a ridge; these feathers are brown tinged with greenish, and so is the upper side of the head: the mantle is brown, and the breast silvery white, as in the other grebes. It is of this in particular that the nest is said by Linnæus to float on the water: he adds, that it lays four or five eggs, and that the female is entirely gray *.

* *Fauna Suecica*, N° 123.

It

It is known in moſt countries of Europe, whether maritime or inland. Edwards received it from Hudſon's Bay *. But its being found in North America is no reaſon why Briſſon ſhould infer that it is the ſame with the *Yaca-pitzahoac* of Fernandez; which indeed appears to be a grebe, but is not ſufficiently characte-rized. With regard to the *Trapazorola* of Geſner, as likewiſe of Briſſon, it is moſt proba-bly a *cheſnut*; at leaſt it is not a horned grebe, ſince Geſner expreſsly mentions its having no creſt.

* We will not heſitate to refer the *Eared Dobchick* of that ſame naturaliſt, notwithſtanding ſome differences in dimenſions, to the Little Horned Grebe.

The BLACK-BREASTED GREBE.

Le Grebe Duc-laart. *Buff.*

SEVENTH SPECIES.

Colymbus Thomenſis.   Gmel.
Podiceps Thomenſis.   Lath.
Colymbus Inſulæ St. Thomæ.   Briſſ.

THIS Grebe is called the *laart duck* in the iſland of St. Thomas, where Father Feu-illée obſerved and deſcribed it. What diſtin-guiſhes it moſt, is a black ſpot in the midſt of

its

its fine white breaft-plate, and the colour of its wings, which is pale rufous. It is as large, he tells us, as a pullet : he remarks alfo, that the point of its bill is flightly curved, a property which belongs equally to the following fpecies *

* Thus defcribed by Briffon : " Above, dull brown; below, " white, variegated with gray fpots; a bright white fpot on either " fide between the bill and the eye; a black fpot on the middle of " the breaft; the wing-quills pale rufous."

The LOUISIANA GREBE.

EIGHTH SPECIES.

Colymbus Ludovicianus. Gmel.
Podiceps Ludovicianus. Lath.

BESIDES that its bill has a gentle curvature at the tip, its breaft is entirely white; its flanks are deeply ftained with brown and black-ifh, and the fore fide of its neck blackifh. It is fmaller than the common grebe.

[A] Specific chara&er of the *Colymbus Ludovicianus :* " Its " head is fmooth and brown, its body brown, its fides ferruginous, " the middle of its breaft white."

The RED-NECKED GREBE.

LE JOUGRIS*. *Buff.*

NINTH SPECIES.

Colymbus Rubricollis.
Colymbus Subcriftatus. } Gmel.
Podiceps Rubricollis. Lath.

THE cheeks and chin are gray; the fore
fide of the neck is rufous, and the upper
furface dark brown. It is nearly as large as
the horned grebe.

* *i. e.* Joues Grifes, or *Gray Cheeks.*

The GREAT GREBE.

TENTH SPECIES.

Colymbus Cayennenfis. Gmel.
Podiceps Cayanus. Lath. Ind.
The Cayenne Grebe. Lath. Syn.

THE epithet *great* is due more to the length
of its neck than to the bulk of its body:
its head is raifed three or four inches higher
than that of the common grebe. Its upper

furface is brown; the fore fide of its body rufous brown, which extends alfo on the flanks, and fhades the white of the breaft. It is found in Cayenne.

From the enumeration we have made, it appears, that the fpecies of Grebes are diffufed through both continents; they feem difperfed from pole to pole. The *Kaarfaak* * and the *Efarokitfok* of the Greenlanders are probably grebes: and in the antarctic regions, M. Bougainville found, at the Malouine iflands, two birds which appear to belong to this genus rather than to the divers †.

[A] Specific character of the *Colymbus Cayennenfis*: " Its head " is fmooth, its throat rufous, its breaft and belly white."

* " The bird which the Greenlanders call *kaarfaak*, expreffing " its cry by that name, is a fort of *colymbus* : according to them, " it foretells rain or fine weather, according as its tone of voice is " hoarfe and rapid, or foft and lengthened out. They alfo call it " the *fummer bird*, becaufe its appearance announces that joyous " feafon. The female lays near pools of frefh water, and it is pre- " tended that fhe is fo much attached to her brood as to fit even " when the place is overflowed." *Hift. Gen. des Voyages, tom.* xix. *p.* 45. The Greenland duck *with a pointed bill, and a tuft on its head*, mentioned by Crantz, appears alfo to be a grebe. See *ibid.* p. 43.

† " In the Malouine iflands there are two fpecies of fmall di- " vers ; the one has its back of an afh-colour, and its belly white; " the feathers of the belly are fo filky, fo brilliant, and fo clofe, that " we took it for the grebe, which furnifhes the materials of pre- " cious muffs: this fpecies is rare. The other, more common, is " entirely brown, having its belly fomewhat lighter than its back ;

" its

" its eyes are like rubies, and their wonderful vivacity is height-
" ened by the contraſt with a circle of white feathers that ſurround
" them, and which has given occaſion to the name of *ſpectacle diver.*
" It has two young, too delicate no doubt to bear the coolneſs of
" the water when they are clothed only with down, for the mother
" then carries them on her back. Theſe two ſpecies have not their
" feet palmated, like other water fowl; their toes are parted, and
" furniſhed on each ſide with a very ſtrong membrane; in this
" ſtate, each toe reſembles a leaf rounded towards the nail, and the
" more ſo, as from the toe lines proceed to terminate in the extre-
" mity of the membranes, and as the whole has the green colour of
" leaves, without much thickneſs."—*Voyage autour du Monde, par
de Bougainville, tom.* ii. *p.* 117, 118.

The CHESNUT

LE CASTAGNEUX. *Buff.*

FIRST SPECIES.

Colymbus Minor. Gmel.
Podiceps Minor. Lath. Ray, and Will.
Colymbus Fluviatilis. Briff.
Mergus Parvus Fluviatilis. Gefner.
The Didapper, Dipper, Dobchick, } Will.
  Small Doucker, Loon, or Arfefoot. }
The Little Grebe. Penn. and Lath.

WE have faid, that the Chefnut is much
fmaller than the other grebes: we may
even add, that, except the ftormy petrel, it is
the leaft of all the fwimming birds. It refembles the petrel alfo, in being clothed with down
inftead of feathers. But its bill, its feet, and all
its body, are exactly like thofe of the grebes:
its colours are nearly the fame, but, as its back
is of a chefnut-brown, it has been termed the
caftagneux. In fome individuals, the fore fide of
the body is gray, and not of a gloffy white; in
others, they are more inclined to blackifh than
to brown on the back; and this variety in the
colours has been remarked by Aldrovandus.
Like the grebes, the Chefnut wants the power

of

THE LITTLE GREBE.

of ftanding and walking on the ground; its legs trail and project behind, and cannot fupport it*: with difficulty it rifes, but when once it has mounted, it flies to a great diftance. It is feen on the rivers the whole winter, at which time it is very fat. Though called the *river grebe*, it is feen alfo on the fea-fhore, where it eats fhrimps and fmelts †, as it likewife feeds on young crabs and fmall fifh in frefh waters. We have found particles of fand in its ftomach; this is mufcular, and lined with a glandulous membrane, thick and inadhefive: the inteftines, as Belon obferves, are very weak; the two legs are attached behind the body by a membrane, which projects when they are extended, and is faftened very near the joint of the tarfus; above the rump, and inftead of a tail, are two fmall pencils of down, which rife each out of its tubercle: it is alfo obferved, that the webs of the toes have a border indented with little fcales ranged regularly.

We conceive the *Tropazorola* of Gefner to be the fame bird: that naturalift fays, that the former appears, after winter, on the lakes of Switzerland.

* Belon.			† *Idem.*

[A] Specific character of the *Colymbus Minor:* "It is fcarlet, "below white and fpotted, its head fmooth."

Q 3

The PHILIPPINE CHESNUT.

SECOND SPECIES.

Colymbus Minor, var. Gmel.

THOUGH not larger than the preceding, it is diftinguifhed by two great ftreaks of rufous, which ftain the cheeks and the fides of the neck, and alfo by a purple tinge fpread on the upper furface: perhaps it is only the fame bird, modified by climate. We might pronounce with more certainty, if the limits which feparate them, or the chain that connects them, were better known:—But who can trace the genealogy of nature's families?

The CIRCLED-BILL CHESNUT.

THIRD SPECIES.

Colymbus Podiceps. Linn. and Gmel.
Podiceps Carolinenfis. Lath. Ind.
Colymbus Fluviatilis Carolinenfis. Briff.
Colymbus Fufcus. Klein.
The Pied-bill Dobchick. Catefby.
The Pied-bill Grebe. Penn. and Lath.

A LITTLE black ring, which encircles the middle of the bill, ferves to diftinguifh this Chefnut. It has alfo a remarkable black fpot

at

at the bafe of the lower mandible; its plumage is entirely brown, deep on the head and neck, light and greenifh on the breaft. It is found on pools of frefh water, in the fettled parts of Carolina.

The St. DOMINGO CHES-NUT.

FOURTH SPECIES.

Colymbus Dominicus. Linn. and Gmel.
Podiceps Dominicus. Lath. Ind.
Colymbus Fluviatilis Dominicenfis. Briff.
The Twopenny Chick. Hughes Barbadoes.
The White-winged Grebe. Lath. Syn.

THIS is fmaller than the European Chef-nut; its length, from the bill to the tail, fcarce feven inches and a half; it is blackifh on the body, and filvery light gray, fpotted with brown, below.

The COOT-GREBE.

NATURE never proceeds by starts: she fills up all the intervals, and connects remote objects, by a chain of intermediate productions. The Coot-Grebe, hitherto unknown, is related to both these *genera* of birds. Its tail is pretty broad and its wings long; all its upper surface is olive brown, and all the fore side of the body is a very fine white; the toes and their webs are barred transversely with black and white or yellow stripes, which produces an agreeable effect. It was sent to us from Cayenne, and is as small as our chesnut.

The DIVERS*.

LE PLONGEONS. *Buff.*

THOUGH many aquatic birds dive even to
the bottom of the water in purfuit of their
prey, the name of *Diver* has been appropriated
to a fmall family, diftinguifhed from the reft by
their ftrait pointed bill, and their three fore
toes connected together by an entire membrane,
which throws a hem along the inner toe, from
which the hind one is parted: their nails are
alfo fmall and pointed; their tail is extremely
fhort, and fcarce vifible; their feet are very flat,
and placed quite behind the body; laftly, their
leg is concealed in the lower belly, a difpofition well adapted for fwimming, but almoft incompatible with walking. In fact, the Divers,
when on land, are obliged, like the grebes, to
ftand erect, and cannot maintain their balance:
but in the water their motions are fo nimble
and prompt, that, the inftant they perceive the
flafh of a gun, they plunge and efcape the ball †.

* The general name of the Diver in Greek, is Αιθυα: in Latin
Mergus: in Hebrew and Perfian *Kaath:* in Arabic *Semag:* in Italian *Mergo, Mergone:* in German *Ducher, Duchent, Taucher.*

† " The Divers of Louifiana are the fame with ours, and when
" they fee the fire of the touch-pan, they dive fo nimbly, that the
" lead cannot hit them; for which reafon they are called *lead-*
" *eaters.*" Le Page Dupratz, *Hift. de la Louifiane,* tom. ii. p. 115.

Accordingly,

Accordingly, expert fowlers faften a bit of pafteboard to their piece, in fuch manner as to leave the aim free, and yet fcreen from the bird the gleam of the priming.

We know five fpecies of Divers, two of which, a greater and a leffer, occur equally on the frefh waters of inland countries, and on the falt flood near the fea-fhore: The three other fpecies feem to be attached wholly to the coafts, particularly in the north.

The GREAT DIVER.

FIRST SPECIES.

Colymbus Immer. Linn. and Gmel.
Mergus Major. Briff.
Colymbus Maximus. Gefn. Aldrov. Ray, Johnft. and Klein.
The Imber Diver. Lath.

THIS Diver is almoft as large and as tall as the goofe. It is known on the lakes of Switzerland; and the name *Fluder*, which it receives on that of Conftance, alludes, according to Gefner's remark, to its laborious motion on land, ftruggling at once with its wings and its feet. It never rifes but from the water; and in that element its motions are as eafy as they are rapid.

THE IMBER.

rapid. It dives to very great depths, and fwims
under water to the diftance of an hundred paces
without emerging to take breath: a portion of
air included in its dilated wind-pipe fupplies its
refpiration during this interval. The fame is
the cafe with other divers and grebes; they
glide through the water freely in all directions;
in it they find their food, their fhelter, their
afylum: when the bird of prey hovers above
them, or the fowler appears on the fhore, they
plunge for fafety. But man, ftill more formidable by his addrefs than by his ftrength, prepares
fnares for them, even in the bottom of their retreat: a net or a line baited with a fmall fifh,
allures the unwary bird; it fwallows death with
the repaft, and perifhes in that element where it
received birth; for its neft is placed on the water
amidft the tall rufhes.

Ariftotle juftly obferves, that the Divers begin to breed in early fpring, and that the gulls
do not breed till the end of that feafon, or the
beginning of fummer *. But Pliny, who often
merely copies the philofopher, has here injudicioufly contradicted him, by employing the
name *Mergus* to fignify an aquatic bird which
neftles in trees †; a property which belongs to

* Hift. Animal. *lib.* v. 9.

† *Mergi & in arboribus pariunt,* lib. x. 32. He likewife confounds
the Diver with certain gulls, when he attributes to it the habit of
devouring the excrements of other birds. *Mergi foliti funt devorare
quæ ceteræ reddunt,* lib. x. 47.

the

the cormorant and some other aquatic birds, but which has no relation to the Divers, since they nestle at the foot of rushes.

Some observers have asserted that the Great Diver was very silent: yet Gesner ascribes to it a singular and loud cry; but probably this is seldom heard.

Willughby seems to admit a variety, in which the back is of an uniform colour; whereas, in the ordinary kind, the upper surface is waved with light gray on a brown gray, and this brown, clouded and dotted with whitish, appears on the upper side of the head and neck, which is besides ornamented below by a half collar tinged with the same colours, terminated with fine white on the breast and the under side of the body.

[A] Specific character of the *Colymbus Immer*: " Its body is " blackish above, waved with white, below entirely white."

The LITTLE DIVER.

SECOND SPECIES.

Colymbus Stellatus. Gmel.
Colymbus Maximus Caudatus. Will.
Mergus Minor. Briff.
The Speckled Loon. Alb.
The Speckled Diver. Penn. and Lath.

THIS Diver refembles the preceding in its colours, and has likewife all the fore fide of its body white: its back, and the upper fide of its neck and head, are blackifh-cinereous, entirely fprinkled with little white drops. The largeft of this kind meafure at the utmoft a foot nine inches from the tip of the bill to the end of the tail, and two feet to the extremity of the toes, and the alar extent is two feet and an half; whereas the preceding fpecies is two feet and an half from the bill to the nails, and four feet acrofs the wings. Their natural habits are nearly the fame.

The Little Divers are feen at all times on our pools, which they do not quit till the ice obliges them to flit to the rivers and brooks of running water: they depart in the night-time, and re-move as little as poffible from their former haunts. It was remarked, even in the time of

Ariftotle,

Ariſtotle, that they did not diſappear in winter*.
That philoſopher alſo ſays, that they lay two or
three eggs : but our ſportſmen make them to lay
three or four, and ſay, that when a perſon approaches the neſt, the mother plunges into the
water, and that the young ones juſt hatched
throw themſelves after her. The ſwimming
and diving of theſe birds are always attended
with noiſe, and with a very quick agitation of
their wings and tail. The motion of their feet
impels them forward, not in the line of their
body, but ſidewiſe in the diagonal. M. Hebert
obſerved this in a captive Diver, which being
held by a long ſtring, took always that direction : it appeared to have loſt nothing of its natural liberty : it was kept on a river, where it
lived by catching ſmall fiſh.

* Hiſt. Animal. *lib.* v. 9.

[A] The ſpeckled Diver lays its eggs near the verge of maritime lakes : they are oval, duſky with ſome black ſpots, and as large
as thoſe of a gooſe.

The SEA-CAT DIVER.

THIRD SPECIES.

THIS Diver, which is very like the little
freſh-water Diver, was ſent to us from the
coaſts of Picardy, which it frequents, particularly

larly in winter, and where it is called by the
fifhers *cat-marin*, becaufe it eats much young
fry. They are often caught in the nets fpread
for the fcoter-ducks, with which they generally
arrive; for they are obferved to retire in fummer, as if they fpent that feafon in more northern countries. Some, however, breed in the
Scilly ifles on the rocks, which they gain by
fpringing from the water, having taken advantage of a fwell: for, like the other Divers, they
can hardly rife from the land *; nor can they
even run but on the waves, which they rapidly
graze in an erect attitude, the lower part of their
body being immerfed.

This bird enters with the tide into the mouths
of rivers. It prefers fmall fmelts, and the
fry of the fturgeon and conger. As it fwims
almoft as faft as other birds fly, and dives as
well as fifh, it has every poffible advantage for
feizing its fugitive prey.

The young ones, lefs dextrous and lefs experienced than the old, fubfift only on fhrimps;
yet both are, at all feafons, extremely fat. M.
Baillon, who has carefully obferved thefe Divers
on the coafts of Picardy, and who has furnifhed

* " I one day found two of thefe Divers, which had been caft
" afhore by the waves; they were lying on the fand, working their
" feet and wings, and crawling with difficulty: I gathered them
" like ftones, yet they were not wounded; and one of them thrown
" up, flew, dived, and played on the water before my eyes." *Obfervation communicated by M. Baillon, of Montreuil-fur-mer.*

I   us

us with thefe details, fubjoins, that the female is
diftinguifhed by being two inches fhorter than
the male, which meafures two feet three inches
from the point of the bill to the end of the
nails, and its alar extent three feet two inches:
the plumage of the young ones, till moulting,
is a fmoky black, without any of the white
fpots fprinkled on the back of the adults.

We fhall refer to this kind, as a variety, a
black-headed Diver, which Briffon makes his
fifth fpecies, and applies to it the names given
by Willughby and Ray, which refer only to the
northern Diver.

It has been remarked, though not with regard to any particular fpecies of Divers, that the
flefh of thefe birds is improved by living, in
Lough Foyle, near Londonderry in Ireland, on a
certain plant, whofe ftalk is foft, and almoft as
fweet, it is faid, as a fugar-cane.

THE NORTHERN DIVER;

The IMBRIM, or GREAT NORTHERN DIVER.

FOURTH SPECIES.

Colymbus Glacialis. Linn. and **Gmel.**
Mergus Major Nævius. Briff.
Colymbus Maximus Stellatus. Sibb.
Colymbus Maximus Caudatus. Ray, Will. and Klein.
The Greateſt Speckled Diver, or Loon. Alb.
The Northern Diver *. Penn. and Lath.

IN the Feroe iflands this Great Diver is called the *Imbrim,* and in the Orkneys *Embergooſe.* It is larger than a goofe, being near three feet from the bill to the nails, and four feet over the wings: it is alfo remarkable by a furrowed collar about the neck, marked by fmall longitudinal ftripes, alternately black and white: the ground on which this belt lies is black, with green reflections on the neck, and violet ones on the head: the mantle is black, entirely fprinkled with white fpeckles; all the under fide of the body is fine white.

* In Norwegian *Bruſen:* in Icelandic *Huubryre,* according to Anderfon, who fays that this bird much refembles the vultur by its bulk and its cries; but this pretended vultur is a merganfer.

This

This Great Diver appears fometimes in England in hard winters *; but at other times it
never leaves the northern feas, and its ufual retreat is among the Orkneys, the Feroe iflands,
on the coafts of Iceland, and near Greenland;
for it is evidently the *Tuglek* of the Greenlanders †.

Some writers of the north, fuch as *Hoierus*,
phyfician at Bergen, have afferted that thefe
birds make their nefts and lay their eggs under
water; which is not even probable ‡: and the
account inferted in the Philofophical Tranfactions §, that the Imbrim hatches her eggs by
carrying them under her wings, appears to me
equally fabulous. All that we can infer from
thefe ftories is, that this bird probably breeds
on fhelves or defart coafts and that no obferver
has yet feen its neft.

* Ray.—We received one that was killed this winter (1780) on
the coaft of Picardy.

† " The Tuglek," fays Crantz, " is a diver of the bulk of a tur-
" key-cock, and of the colour of a ftare; its belly white, and its
" back fprinkled with white; its bill is ftraight and pointed, an inch
" thick, and four inches long; its length from the head to the tail
" is two feet, and its alar extent two feet." *Hift. Gen. des Voyages,*
tom. xix. *p.* 45.

‡ Klein juftly refufes to credit the report.

§ Nº 473. p. 61.

[A] Specific charaƈter of the *Colymbus Glacialis:* " Its head
" and neck are dark violet; a white interrupted bar on its throat
" and neck." In the northern regions, every pair of thefe birds
occupy a lake, and breed on the fmall iflets. The young defend
themfelves courageoufly with their bills. The Greenlanders ufe
the fkins for cloathing, and the Efquimaux deck their heads with
the feathers.

The LUMME, or LITTLE DIVER
of the NORTHERN SEA.

FIFTH SPECIES.

Colymbus Arcticus. Linn, Gmel. Sibb. and Will.
Mergus Gutture Nigro. Briff.
Mergus Arcticus fimpliciter. Klein.
The Black-throated Diver *. Edw. Penn. and Lath.

LUMME or *Loom* in Lapponic fignifies *lame*, alluding to the tottering pace of this bird when on land: it feldom however comes afhore, but fwims almoft conftantly, and breeds at the verge of the flood on defert coafts. Few have feen its neft, and the people of Iceland fay that it hatches its eggs under its wings in open fea †; which is not more probable than the incubation of the imbrim under water.

The Lumme is fmaller than the imbrim, and about the bulk of a duck; its back is black fprinkled with little white fquares; the throat is black, and alfo the fore fide of the head, of which the upper fide is covered with gray feathers; the top of the neck is clothed with fimi-

* In Swedifh and Lapponic *Loom* or *Lom*: in Greenlandic *Apa*, according to Anderfon, and *Moquo*, according to Edwards.
† Anderfon.

lar

lar gray feathers, and ornamented behind by a
long patch clouded with black, varying with
violet and green : a thick down, like that of the
fwan, covers all the fkin; and the Laplanders
make winter bonnets of thefe fine furs *.

It appears that thefe Divers fcarcely ever quit
the northern feas; though, according to Klein,
they vifit from time to time the coafts of the
Baltic, and are well known through the whole
of Sweden †.   Their principal abode is on the
fhores of Norway, Iceland, and Greenland:
thefe they frequent the whole fummer, and
there breed their young, which they rear with
fingular care and folicitude.   On this fubject,
Anderfon gives details which would be interefting, were they all accurate.   He fays that they
lay only two eggs, and that as foon as a young
Lumme is able to quit the neft, the parents lead
it to the water, the one flying always above it
to keep off the bird of prey, and the other below to receive it in cafe it fhould fall; and that
if notwithftanding their affiftance, the neftling
fall to the ground, the parents rufh after it, and
rather than forfake it, they fuffer themfelves to
be caught by men or eaten by foxes, which ever
watch thofe opportunities, and which, in thofe
bleak frozen regions, are conftrained to turn all
their fagacity and wiles againft the birds.   This

* *Fauna Suecica,* and *Hift. Gen. des Voyages, tom.* xv. *p.* 309.
† *Fauna Suecica.*

author

author adds, that when the Lummes have once reached the fea with their young, they return no more to land. He affirms even that the old ones which have accidentally loft their family, or are paft breeding, never revifit the fhore, but fwim always in flocks of fixty or a hundred. " If we throw a young one into the fea before a " flock of Lummes, they will all gather round " it, and ftrive to attend it ; nay they will fight " about it till the victor leads it off : but if the " mother happen to intervene, the quarrel im- " mediately ceafes, and the infant is configned " to her care."

On the approach of winter, thefe birds retire, and appear not again until the fpring. Anderfon conjectures, that, fhaping their courfe between the eaft and the weft, they arrive in America : and Fdwards in fact admits that this fpecies is common both to the northern feas of that continent and of Europe. We might add thofe of Afia; for the red-throated diver brought from Siberia, and reprefented under that name in our *Pl. Enl.* is exactly the fame with that of Edwards, *pl.* 97, which this naturalift gives as the female Lumme from the unimpeached veracity of his correfpondent Ifham, a good obferver, who fent both cock and hen from Greenland.

When the Lummes vifit the coafts of Norway, their different cries are interpreted by the inhabitants to prefage fine weather or
rain.

rain *. This is probably the reafon why they fpare the lives of thefe birds, and are concerned to find them taken in their nets †.

Linnæus diftinguifhes a variety in this fpecies ‡, and fays with Wormius, that the Lumme makes its neft flat on the beach at fea-mark: on that head, Anderfon contradicts himfelf. The Spitzbergen *Lumb* of Martens appears, according to Ray's obfervation, to be different from the Lummes of Greenland and Iceland, fince *its bill is hooked:* yet its attachment to its young, and the manner in which it leads them to the fea, defending them from the bird of prey, fhow a great analogy to thefe birds in its natural habits §. With regard to the *Loms* of the navigator

---

* " When it forefees abundant rains, fearing that its neft will " be overflowed, it ftrikes the air with a querulous found; on the " contrary, when it expects fine weather, it chears its young with " loud calls and another more grateful found."—*Wormius.*

† *Idem.*

‡ " A variety, whofe head and fides of its neck are cinereous; " the hind part of its neck marked with fmall black and white lines; " its back brown, without the white dots; its breaft fpotted before " with cinereous and white." *Fauna Suecica,* N° 121.

§ " The bill of the *Lumb* refembles much that of the diver pi- " geon, except that it is fomewhat harder and more hooked. " This bird is as large as a middling duck . . . the young are " commonly feen near the old ones, which inftruct them to fwim " and dive; the old tranfport their brood from the rocks into the " water. by taking them in their bill; the burgomafter, which is a " bird of prey, feeks to carry them off . . . but thefe birds are " fo attached to their young, that they will rather be killed than " forfake them, and they defend them as a hen does her chickens; " they cover them as they fwim . . . they fly in large flocks, and " their

gator Barentz, they may be the fame with our
Lummes, which might eafily frequent Nova
Zembla *.

" their wings are fhaped like thofe of fwallows; in flying they ex-
" ercife thefe extremely . . . their cry is very difagreeable, and
" nearly like that of a raven, nor is there any bird that cries more
" than this, unlefs perhaps the *winter rotger.*"—*Recueil des Voyages
du Nord, tom.* ii. *p.* 95.

* " The name of *Loms,* which Barentz gives to this bay (in the
" Icy Sea, under Nova Zembla) was taken from a fpecies which
" abounds there, and which, according to the fignification of the
" Dutch word, are exceffively unwieldy; their body is fo large in
" comparifon of their wings, that one is furprized that they can
" raife fo cumbrous a weight. Thefe birds make their nefts on
" craggy mountains, and cover only one egg at a time. The fight
" of men difturbs them fo little that we may take one in its neft,
" and yet the reft will not fly away, or even fhift their place."—*Hift.
Gen. des Voyages, tom.* xv. *p.* 104.

[A] Specific charafter of the *Colymbus Arfticus:* " Its head is
" hoary, the under fide of its neck dark violet; a white inter-
" rupted bar."

The MERGANSER.

LE HARLE *. *Buff.*

FIRST SPECIES.

Mergus Merganser. Linn. and Gmel.
Merganser. Gesn. Ald. Johnst. Will. Sibb. Briss. &c.
The Goosander, male; *Dun-diver, or Sparling-fowl*, female. Will.

" THIS bird," says Belon, " commits as " much havock in a pool as a beaver;" and hence, he adds, it was termed *bievre*. But the old naturalist was here deceived with the vulgar, for the beaver does not eat fish; and the *otter* is the animal to which this *icthyophagous* bird should be compared.

The Merganser is of a middle size, between the duck and the goose: but in its stature, its plumage, and its short flight, is more allied to the duck. Its name, *diver-goose, (mergus-anser)* seems to have been formed by Gesner injudiciously; for the resemblance of its bill to that of

* In German *Meer-rach, Weltch-eent :* on the lake of Constance *Gan* or *Ganner :* on the lake Maggiore *Garguney :* in Polish *Kruk morski :* in Norwegian *Fisk and, Mort-and :* in Swedish *Wrakfogel, Kjorkfogel, Ard, Skraka :* in Danish *Skallesluger :* in Icelandic *Skior-and :* in the language of Greenland *Peksok.*

the

THE GOOSANDER .

the diver, on which that appellation refts, is very imperfect. The bill is nearly cylindrical, and ftrait to the point, like that of the diver; but differs inafmuch as the point is bent in the fafhion of a crooked nail, with a hard, horny fubftance; it differs alfo becaufe the edges are befet with indentings reflected backwards: the tongue is rough, with hard *papillæ* turned backwards like the indentings on the bill, which ferve to hold the flippery fifh, and even to draw it into the throat of the bird: accordingly, with a gluttonous voracity, it fwallows fifh much larger than can enter entire into its ftomach: the head firft lodges in the œfophagus, and is digefted before the body can defcend.

The Merganfer fwims with all its body fubmerged, and only its head out of the water *: it dives deep, remains long under water, and traverfes a great fpace before it again appears. Though its wings are fhort, it flies rapidly, and ofteneft it fhoots above the furface of the water †: it then appears almoft entirely white, and is therefore denominated *harle blanc* in fome parts of France, as in Brie, where however it is rare. Yet the fore fide of its body is wafhed with pale yellow: the upper fide of the neck and all the head are black, changing by reflections into green; and the feathers, which are flender, filky, long, and briftled up from the nape

* Aldrovandus and Wormius.
† Rzyczynfki.

X   to

to the front, augment much the bulk of the
head : the back confifts of three colours, black
on the top and on the great coverts of the wings,
white on the middle ones and moft of the coverts, and handfomely fringed with gray upon
white at the rump : the tail is gray : the eyes,
the feet, and part of the bill, are red.

The Merganfer, we have feen, is a very beautiful bird; but its flefh is dry and unpleafant
food *. The form of its body is broad and fenfibly flattened on the back. Its wind-pipe is
obferved to have three fwellings, the laft of
which, near the bifurcation, includes a bony la-
byrinth † : this apparatus contains the air which
the bird refpires under water ‡. Belon fays
alfo, that he remarked that the tail of the Merganfer was often rumpled and turned up at the
end, and that it perches and builds its neft, like
the cormorant, on trees or rocks : but Aldrovandus afferts, with more probability, that it
breeds on the fhore, and never quits the water.
We have not been able to afcertain this fact;
thefe birds appear only at diftant intervals in
France, and from all the accounts which we
have received, we can only gather that they occur in different places, and always in win-

---

* Belon relates the vulgar proverb, *He who would regale the de-
vil, might ferve him with merganfer and cormorant.*

† Willughby.

‡ Belon.

ter.

ter *. In Switzerland their appearance on the
lakes is fuppofed to forebode a fevere winter:
and though they muft be known on the Loire,
fince there, according to Belon, they received
the name of *harle* or *herle*, that obferver himfelf intimates, that they retire in winter to more
fouthern climates, for he faw them entering
from the north into Egypt; yet he fays, that in
every other feafon except winter they are found
on the Nile, which is difficult to reconcile.

The Merganfers are not more common in
England than in France †; yet they penetrate
into Norway ‡, Iceland §, and perhaps ftill farther north. The *Geir-fogel* of the Icelanders,
which Anderfon improperly calls a vulture, is a
Merganfer; at leaft if its voracity may entitle
it to the appellation of fea vulture. But it feems
thefe birds do not conftantly refide on the coaft
of Iceland; fince every time they arrive, the inhabitants expect fome great event.

The female Merganfer is uniformly fmaller
confiderably than the male: it differs alfo, like
moft of the water birds, by its colours; its head

* Merganfer killed the 15th of February 1778, at Montbard, on
a pool, where it had been feen for feveral days.—Merganfer killed
near Croifie, on the falt marfhes.—*Letter of M. de Querhoënt, of
the 13th of February.*—Merganfer killed at Bourbon-lancy, and fent
to M. Hebert in March 1774.

† Charleton.
‡ Muller.
§ Wormius, Charleton.

is

is rufous, and its mantle gray *. Briſſon makes
it his ſeventh ſpecies.

* It is the female which Belon ſtyles the *beaver*. Linnæus, in
the twelfth edition of his *Syſtema Naturæ*, under the name *Mergus-
Caſtor*.

[A] Specific character of the *Mergus-Merganſer* : " Its creſt
" longitudinal and ſomewhat erect; its breaſt whitiſh and ſpotleſs;
" its tail-quills cinereous, their ſhaft blackiſh." Linnæus ſays,
that theſe birds breed ſometimes on trees and ſometimes between
ſtones, and lay fourteen eggs. They paſs the whole year in the
Orknies, yet never appear in England except in hard winters.
They are found not only in the north of Europe, but in the greater
part of North America.

The CRESTED MERGANSER.

Le Harle Huppe . *Buff.*

SECOND SPECIES.

Mergus-Serrator. Linn. and Gmel.
Serrator Cirrhatus. Klein.
Mergus Criſtatus. Briſſ.
The Serula. Will.
The Leſſer Dun Diver. Penn.
The Leſſer Toothed Diver. Morton.
The Red-breaſted Merganſer. Lath.

THE preceding ſpecies had only a tuft; this
is adorned with a diſtinct and well-formed
creſt, conſiſting of ſlender, long plumules, directed backwards from the occiput : it is about
the

the size of a duck : its head and the top of its neck are of a violet black, changing into gold green : the breast is rufous variegated with white; the back is black; the rump and the flanks are striped in zig-zags with brown and light gray; the wing is variegated with black, with brown, with white, and with cinereous, on both sides of the breast, near the shoulders, there are pretty long white feathers edged with black, which cover the pinion when the wing is closed; the bill and feet are red. The female is distinguished from the male by its head being of a duller rufous, its back gray, and all the fore side of its body white, faintly tinged with fulvous on the breast.

According to Willughby, this species is very common on the lagoons of Venice; and since Muller affirms that it is found in Denmark and Norway, and Linnæus, that it also inhabits Lapland *, it probably frequents the intermediate countries. In fact, Schwenckfeld assures us, that this bird passes into Silesia, where it is seen in the beginning of winter on the pools among the mountains. Salerne says that it is very common on the Loire ; but from his manner of speaking of it, he seems to have observed it very inattentively.

* The *Knipa* of Schoeffer, *Fauna Suecica.*

[A] Specific character of the Crested Merganser, *Mergus-Serrator :* " Its crest is hanging; its breast tawny and variegated;
" its

" its neck white; its tail-quills brown, variegated with cinereous."
In Iceland thefe birds are called *Vatus-ond*. They appear in great
flocks during the fummer in Hudfon's Bay, and on the Siberian
lakes.

The PIETTE, or LITTLE CRESTED MERGANSER.

THIRD SPECIES.

Mergus Albellus. Linn. and Gmel.
Mergus Varius. Gefner.
Mergus Rheni. Aldrov. Ray, &c.
The Smew, or White Nun. Penn. and Lath.

THIS is a handfome little Merganfer with a
pied plumage : it is fometimes called the
Nun (Religieufe) no doubt becaufe of the neat-
nefs of its fine white robe, its black mantle, its
head hooded with white unwebbed feathers,
difpofed like a chin-piece, and raifed in the form
of a band, which interfects behind a little veil
lappet of a dull green-violet : a black half collar
on the top of the neck compleats the modeft
and elegant apparel of this little winged Nun.
It is alfo well known by the denomination
Piette * on the rivers Are and Somme in Picardy, where is not a peafant, fays Belon, but
knows its name.   It is rather larger than the

* From *pietter, to trip lightly.*

garganey,

THE SMEW MERGANSER.

garganey, but fmaller than the morillon; its bill
is black, and its feet of a lead gray: the extent of black and white on its plumage is very
fubject to vary, infomuch that it is fometimes
almoft all white. The female is not fo beautiful as the male; it has no creft; its head is
rufous, and its mantle is gray.

[A] Specific character of the Smew, *Mergus-Albellus*: " Its
" creft is hanging; the back of its head black; its body white;
" its back and temples black; its wings variegated."

The MANTLED MERGANSER.

FOURTH SPECIES.

Mergus Serrator, var. 1. Gmel.
Mergus Niger. Gefner and Johnfton.
Merganfer Leucomelanus. } Briffon.
Merganfer Niger.
Anas Longiroftra   tertia. } Schwenckfeld.
———————— *fexta.*

WE rank thefe birds together, becaufe they
differ lefs than the male and female in this
genus; efpecially as they are nearly of the fame
fize. Belon, who has defcribed one under the
name of *tiers (third)* fays that it was fo called
as being intermediate, or *the third between the
duck and the morillon,* and that its wings imitate,
by their motley colours, the variety of the morillon's

rillon's wings: he was miftaken however in joining his *tiers* with this bird, fince its bill is entirely different from that of the morillon; and its bulk approaches more to that of the duck. This defcription exactly fuits then the *Mergus Leucomelanus (black and white merganfer)* of Briffon; it alfo agrees with his *Mergus Niger*, (or *black merganfer)* except that the neck of this laft has a bay caft, and that its tail is black: the bill and feet of both are red. Schwenck-feld fays, that the former are feldom feen in Silefia; but he does not by that expreffion in-finuate that the latter is more common there, while he remarks that fome of thefe appear on the rivers in March on the breaking up of the ice.

The STELLATED MERGANSER.

LE HARLE ETOILE, *Buff.*

FIFTH SPECIES.

Mergus Minutus. Linn.
Merganfer Stellatus. Briff. and Klein.
Mergus Albus. Gefner and Johnfton.
Mergus Glacialis. Aldrov. Will. and Charl.
The Weefel Coot. Albin.
The Minute Lough Diver. Penn.

THE great difference between the male and female in this genus occafions much con-fufion in the nomenclature: and we ftrongly
fufpect

fufpect that if the Stellated Merganfer were better defcribed and better known, it would be
found to be the female of fome of the foregoing
fpecies. Willughby was of this opinion, and
regarded it as the female of the mantled Merganfer; and indeed it has the peculiar property
of that bird, being found fometimes entirely
white. Briffon gives it the epithet *ftellated*,
from a white fpot, figured like a ftar, which is
placed, he fays, below a black fpot that furrounds the eyes. The upper fide of the head is
bay colour, the mantle blackifh brown, all the
fore fide of the body is white, and the wing is
partly white, partly black; the bill is black, or
lead-coloured, as in the mantled Merganfer;
and thefe two birds are nearly of the fame fize.
Gefner fays, that this Merganfer is called in
Switzerland the *Ice Duck*, becaufe it does not
appear on the lakes till hard froft fets in.

[A] Specific character of the *Mergus Minutus*: " Its head is
" fmooth and gray, a black bar on its eye, a white fpot under the
" eye." In winter, thefe birds vifit the fhores of our ifland, from
the northern regions.

The CROWNED MER-GANSER.

SIXTH SPECIES.

Mergus Cucullatus. Linn. and Gmel.
Merganfer Virginianus Criftatus. Briff.
Serrator Cucullatus. Klein.
The Wind-Bird. Will.
The Round-crefted Duck. Catefby and Edw.
The Hooded Merganfer. Penn. and Lath.

THIS Merganfer, which is found in Virginia, is very remarkable for a fine edged
crown on its head, black in the circumference,
and white in the middle, formed of feathers elevated to a difk; which has a fine effect, but appears to advantage only in the living bird. Its
breaft and belly are white; the bill, the face, the
neck, and the back, are black; the quills of the
tail and wings are brown; the innermoft in the
wings are black, and marked with a white ftreak.
This bird is nearly as large as a duck: the female is entirely brown, and its creft is fmaller
than that of the male. Fernandez has defcribed
both under the Mexican name *Ecatototl*, with
the epithet *wind-bird*, without mentioning the
reafon. Thefe birds are found in Mexico and
Carolina, as well as in Virginia, and haunt the
rivers and pools.

[A] Specific character of the *Mergus Cucullatus*: "Its creft is
" ball-fhaped and white on both fides; its body brown above, and
" white below." It winters in Virginia and Carolina.

THE GREAT WHITE PELICAN.

The PELICAN.

LE PELICAN. *Buff*.

Pelecanus Onocrotalus. Linn. and Gmel.
Onocrotalus. Gefner. Aldrov. Will. Johnft. Briff. &c.
The White Pelican. Edw. Penn. and Lath.

THE Pelican * is more interefting to the naturalift by its greater ftature and the large fac under its bill, than by the fabulous celebrity of its name, facred among the religious emblems of ignorant nations. It has been employed to re-

---

* In Greek Πελεκανος, Πελεκινος, Πελεχινος, in different authors, from Πελεκυς, *a hatchet*, on account of its broad bill: it had alfo the name ΟνοκρόΙαλος, from ονος an afs, and κρόΙολον a rattle, becaufe of the gurgling in its throat. The Romans adopted that term; but, according to Verrius Flaccus and Feftus, they anciently called it *Truo*. In Hebrew it was denominated *Kakik*: in Chaldean *Catha*: in Arabic *Kuk* and *Alhaufal*, meaning *gullet*: in Perfian *Kik Tacab*, (which fignifies *water-carrier*) or *Mifo (fheep*, on account of its bulk): in Egyptian *Begas* or *Gemel-el bahr (water camel)*: in Turkifh *Sackagufch*: in the old Vandal language *Bukriez*: in Spanifh *Groto*: in Italian *Agrotto*: at Rome *Truo*; and near Sienna and Mantua *Agrotti*: in the Alps of Savoy *Goettreufe*, becaufe its bag refembles the *goitres* to which the mountaineers are fubject: in German *Meergans, Schnee-gans (fea-goofe, fnow-goofe)*: in Auftria *Ohne-Vogel* (the *awme*, or *tierce-bird*): in Polifh *Bak, Bak Cudzoziemfki*: in Ruffia *Baba*: in modern Greek *Toubano*: in the French Weft India iflands *Grand Gofier (great gullet)*: in Mexico *Atototl*; and by the Spanifh fettlers *Alcatraz*: in the Philippine iflands *Pagala*: by the negroes of Guinea *Pokko*: by the Siamefe *Noktho*: in old French *Livane*.

present

prefent maternal tendernefs, tearing its breaft to nourifh its languifhing family with its blood. This tale, which the Egyptians had before related of the vulture *, cannot apply to the Pelican, which lives in abundance †, and even enjoys an advantage over the other pifcivorous birds, being provided with a bag for ftoring its provifions.

The Pelican equals, or even furpaffes, the bulk of the fwan ‡, and would be the largeft of all the aquatic birds §, were not the albatrofs thicker, and the flamingo much taller on its legs. Thofe of the Pelican, on the contrary, are very low; but its wings are fo broad as to extend eleven or twelve feet ||. It therefore fupports itfelf eafily, and for a length of time, in the air: it balances itfelf with alertnefs, and never changes its place but to dart directly downwards on its prey, which cannot efcape; for the violence of the

* Horus Apollo.

† St. Auguftine and St. Jerome feem to be the authors of the application of this fable, originally Egyptian, to the Pelican.

‡ Edwards reckons the one which he defcribes twice as large as the fwan. Ellis fpeaks of one more than double the bulk of a large fwan.

§ " I fat out on the fecond of October for the ifland of Griel by " this channel, which is parallel to the main branch of the Niger . . . " it was entirely covered with Pelicans, which were failing gravely " like fwans on the water; they are indifputably, after the oftrich, " the largeft birds of the country." Adanfon, *Voyage au Senegal*, *p.* 136.

|| The Pelicans defcribed by the academicians had eleven feet of alar extent, which, as they remark, is double that of the fwans and of the eagles.

dafh,

dafh, and its wide-fpread wings, which ftrike
and cover the furface of the water, make it boil
and whirl *, and at the fame time ftun the fifh,
and deprive it of the power of flight †. Such
is their mode of fifhing when alone; but in large
flocks they vary their manœuvres, and act in
concert: they range themfelves in a line, and
fwim in company, forming a large circle, which
they contract by degrees to inclofe the fifhes ‡,
and they fhare the capture at their convenience.

These birds fpend in fifhing the hours of the
morning and evening, when the finny tribe are
moft in motion; and chufe the places where
they are moft plentiful. It is amufing to behold
them fweeping the water, rifing a few fathoms
above it, falling with their neck extended and
their fac half full; then afcending with effort to
drop again §, and continuing this exertion till their
wide bag is entirely filled. Now they retire to
eat, and digeft at leifure on fome cliffs, where
they remain tranquil and drowfy till evening ‖.

It appears to me, that this inftinct of the Pelican, of not fwallowing its prey at firft, but
collecting a provifion, might be turned to account; and that, like the cormorant, it might be

* *Petr. Martyr.* Nov. Orb. decad. i. lib. 6.
† Labat and Dutertre.
‡ Adanfon, *Voyage au Senegal,* p. 136.
§ Nieremberg, *Hift. Nat. lib. x. p.* 223.
‖ Labat and Dutertre.

made

made a domestic fisher: indeed travellers affirm, that the Chinese have actually succeeded *. Labat relates, that the savages trained a Pelican, which they dispatched in the morning, after having stained it red with *rocou* †, and that it returned in the evening to their hut with its sac full of fish, which they made it to disgorge.

This bird must be an excellent swimmer; its feet are completely webbed, its four toes being connected by a single piece of membrane: this skin and the feet are red or yellow, according to the age ‡: and it seems, as the Pelican grows old, to assume that fine, soft, and almost transparent rosy tint, which gives its white plumage the lustre of a varnish.

The feathers on its neck are only a short down; those on the nape are longer, and form a sort of tuft; its head is flat at the sides; its eyes are small, and placed in two broad naked cheeks; its tail is composed of eighteen quills; the colours of its bill are yellow and pale orange on a gray ground, with streaks of bright red on the middle and near the extremity; this bill is flattened above like a broad blade, with a longi-

---

* See Voyage de Pirard, *Paris*, 1619. *tom.* i. *p.* 376. But Pirard is mistaken when he thinks that this bird is peculiar to China.

† Probably the same with the *puccoon*, employed by the Indians to heighten their copper complexion, and held in great estimation among them. It is the root of the *Sanguinaria Canadensis*, a low herbaceous plant, which bears a fine white flower in the spring, and is scattered profusely in the American forests.—*T.*

‡ Aldrovandus.

tudinal

tudinal ridge, terminating in a hook; the infide
of this blade, which makes the upper mandible,
has five protuberant wrinkles, of which the two
outer form the cutting edges; the lower mandible confifts only of two flexible branches, which
accommodate themfelves to the extenfion of the
membranous pouch attached to them, and which
hangs below as a fac in fafhion of a bow-net.
This pouch can hold more than twenty quarts of
liquid* : it is fo wide and fo long, that a perfon
may put his foot in it †, or thruft his arm as far
the elbow ‡. Ellis fays, that he has feen a man
cover his head with it §; which will not, however, make us credit what Sanctius ‖ fays, that
one of thefe birds dropt in the air a negro child,
which it had carried up in its fac.

This large bird appears fufceptible of fome
education, and even of a certain cheerfulnefs, notwithftanding its weight ¶. It has nothing fa-

* " The length of the bill of the Pelican which I meafured was
" more than a foot and half, and its fac contained near twenty-
" two pints of water." (Equal to forty-four Englifh wine pints. *T.*)
Adanfon, *Voyage au Senegal, p.* 136.

† Belon.

‡ Gefner.

§ This is commonly exhibited by the keepers of wild beafts in
London.—*T.*

‖ In Aldrovandus, *tom.* iii. *p.* 50.

¶ Belon.—" It was diverting to fee, when we fet upon it the
" boys or our dogs, how admirably it defended itfelf, rufhing with
" impetuofity on its antagonifts, and ftriking them neatly with its
" bill, which they equally repaid; fo that it looked as if they were
" beating two fticks againft each other, or playing with clatter-
" bones." *Voyage en Guinée, par Guillaume Bofman, Utrecht,* 1705.
Lettre xv.

vage,

vage, but foon becomes familiar with man *.
Belon faw one in the ifle of Rhodes, which
walked freely through the town; and Culmann,
in Gefner, relates the noted ftory of the Pelican
which followed the emperor Maximilian, flying
over the head of his army when on a march,
and rifing fometimes fo high as to feem like a
fwallow, though it meafured fifteen Rhenifh feet
acrofs the wings.

This vaft power of flight would be aftonifhing in a bird that weighs twenty-four or twenty-
five pounds, were it not wonderfully affifted by
the great quantity of air with which its body is
inflated, and alfo by the lightnefs of its fkeleton,
which exceeds not a pound and half; its bones
are fo thin, that they are fomewhat tranfparent,
and Aldrovandus afferts that they have no marrow. It is no doubt owing to the nature of
thefe folid parts, which are flow in offifying, that
the Pelican enjoys its great longevity † : even
in captivity it has been obferved to live longer
than moft other birds ‡.

* Rzaczynfki fpeaks of a pelican kept fourteen years at the court
of Bavaria, which was very fond of company, and feemed to take
fingular pleafure in hearing mufic. *Auctuar. p.* 399.

† Turner fpeaks of a tame Pelican that lived fifty years. The
one mentioned by Culmann attained the age of fourfcore; and in
its latter years it was maintained by order of the emperor, at the ex-
pence of four crowns a day.

‡ Of a great number of Pelicans kept in the menagerie at Ver-
failles, none have died in the fpace of twelve years; yet during that
time fome of every other fpecies of animal has died. *Memoires de
l'Academie des Sciences, p.* 191.

The

The Pelican, though not entirely foreign, is
very rare in our climates, efpecially in the interior
provinces. We have in our cabinet the bodies
of two which were killed, the one in Dauphiné,
and the other on the Saone *. Gefner fpeaks
of one that was taken on the lake of Zurich, and
was regarded as an unknown bird †. It is not
common in the north of Germany ‡, though
great numbers occur in the fouthern provinces
watered by the Danube § : this noble river was
an ancient haunt of thefe birds; for Ariftotle,
ranging the Pelicans with fome gregarious kinds,
the crane and the fwan, fays, that they depart
from the Strymon, and waiting for each other
at the paffage of the mountains, they all alight
together, and neftle on the banks of the Danube ‖.
Thefe ftreams, therefore, feem to bound the
countries where their flocks advance from north
to fouth in our continent: and Pliny muft have
been ignorant of this route, when he reprefented
them as coming from the northern extremity of
Gaul ¶ : for they are ftrangers there, and ftill
more in Sweden and the arctic tracts, at leaft if
we judge from the filence of their naturalifts **;

* M. de Piolenc fent us one which he had killed in a marfh near
Arles, and M. Lottinger another from a pool between Dieuze and
Sarreburg.
† Aldrovandus, *tom* iii. *p.* 51.
‡ Schwenckfeld relates, that one was caught in 1585, at Breflaw.
§ Rzaczynfki.
‖ Hift. Animal. *lib.* viii. 12.
¶ Hift. Nat. *lib.* x.
** Linnæus, Muller, Brunnich.

the account which Olaus Magnus gives of the
ancient *onocrotalus* being only an ill-digefted compilation. Nor does it feem to frequent England, fince the authors of the Britifh Zoology
do not infert it in their work; and Charleton relates, that in his time there were Pelicans
in Windfor Park, which had been fent from
Ruffia. In fact, they are found, and even pretty
often, in Red Ruffia, and in Lithuania, as well
as in Volhinia, in Podolia, and in Pokutia, as
Rzaczynfki teftifies: but they extend not to the
moft northern parts of Mufcovy, as Ellis pretends. In general, thefe birds feem to affect
more the warm than the cold climates. One of
the largeft fize, weighing twenty-five pounds,
was killed in the ifland of Majorca, near the
bay of Alcudia, in June 1773*. They appear
regularly every year on the lakes of Mantua and
Orbitello; and from a paffage of Martial we
may infer that they were common in the territory of Ravenna†. They are found alfo in
Afia Minor ‡, in Greece §, and in many parts of
the

* *Journal Hiftorique & Politique*, 20 *Juillet* 1773.
† *Turpe Ravennatis guttur onocrotali.*
‡ " *Onocrotales* feed in a lake which is above the city of An-
" tioch." *Belon.*
§ " We killed with ftones (near Patras) one of thofe large birds
" which we call *pelicans*, the Latins *onocrotali*, and the modern Greeks
" *taubano.* I know not whether the cold hindered it from rifing:
" it had a pouch under its bill, into which we poured more than fif-
" teen quarts of water. The Greeks fay that it carries water to
" the mountains for its young. It is very common in thofe parts, as
" well

the Mediterranean and the Propontis. Belon even obſerved at ſea their paſſage between Rhodes and Alexandria; they flew in bodies from north to ſouth, ſhaping their courſe towards Egypt: and the ſame traveller enjoyed a ſecond time this ſight, near the confines of Arabia and Paleſtine. Laſtly, voyagers tell us, that the lakes of Judea and of Egypt, the banks of the Nile in winter, and thoſe of the Strymon in ſummer, ſeen from the heights, appear whitened by the multitude of Pelicans which cover them.

When we collect the teſtimonies of the various navigators, we ſee that the Pelicans inhabit all the ſouthern countries of our continent, and that they occur, with little difference, and in ſtill greater numbers, in the correſponding parallels in the new world. They are very common in Africa, on the ſides of the Senegal, and of the Gambia, where the negroes call them *pokko* [*]: the great tongue of land, which bars the mouth of the firſt of theſe rivers, is filled with them [†]. They are found likewiſe at Loango, and on the coaſts of Angola [‡], of Sierra Leona [§], and of Guinea[||]: on the bay of Saldana they are intermingled

----

" well as on the coaſt of Smyrna." *Wheeler and Spon's Travels into Dalmatia.*

 [*] Moore; *Hiſt. Gen. des Voy. tom.* iii. *p.* 304.—Voyage de le Maire aux Canaries, *Paris,* 1695, *p.* 104.

 [†] Brue; *Hiſt. Gen. des Voy. tom.* ii. *p.* 488.

 [‡] Pigafetta.

 [§] Finch; *Hiſt. Gen. des Voy. tom.* iii. *p.* 226.

 [||] Voyage de Degenes, *Paris,* 1698, *p.* 41.

with a multitude of birds, which feem, on that
fhore, to fill the air and the fea *. They occur
at Madagafcar †, at Siam ‡, in China §, at the
ifles of Sunda ||, and at the Philippines ¶, efpe-
cially on the fifheries of the great lake of Manilla **. They are fometimes met with at
fea ††: And laftly, they have been feen on the
remote lands in the Indian ocean, as at New
Holland ‡‡, where Captain Cook fays they are
extremely large.

In America, the Pelicans are found from the
Antilles §§ and *Terra Firma* ||||, the ifthmus of
Panama ¶¶, and the bay of Campeachy ***, as far
as Louifiana †††, and the country adjoining to

* Downton; *Hift. Gen. des Voy. tom.* ii. *p.* 46.
† Cauche; *Paris* 1651, *p.* 136.
‡ Tachard; *Hift. Gen. des Voy. tom.* ix. *p.* 311.
§ Pirard.
|| Pifon.
¶ Philofophical Tranfactions, N° 285.
** Sonnerat.
†† " On the 13th of December, after having paffed the Tropic,
" many birds vifited us; there were a great number of Pelicans
" *(grand gofiers)*." *Voyage de le Guat, Amfterdam,* 1708, *tom.* i.
p. 97.
‡‡ *Hift. Gen. des Voy. tom.* xi. *p.* 221.
§§ Dutertre, Labat, Sloane. " In 1656, in the month of Sep-
" tember, there was a great mortality among thefe birds, particu-
" larly the young ones; for all the coafts of the iflands of St. Alou-
" fia, of St. Vincent, of Becouya, and of all the Grenadines, were
" ftrewed with the dead carcafes." Dutertre, *Hift. Gen. des Antilles, tom.* ii. *p.* 271.
|||| Oviedo.
¶¶ Wafer.
*** Dampier.
††† *Hift. Gen. des Voy. tom.* xiv. *p.* 456.

Hudfon's

Hudſon's Bay *. They are ſeen alſo on the inhabited iſles and inlets near St. Domingo †; and
in greater numbers on thoſe ſmall iſles clothed
with the fineſt verdure, which lie in the vicinity
of Guadaloupe, and which ſeem to be occupied
as the retreat of different ſpecies of birds: one
of theſe iſles has even been called *the iſle of Pelicans (l'île aux grand-goſiers ‡)*. They augment alſo the flocks of birds which inhabit the
iſland of *Aves* §; the coaſt of the Sambales,
which abounds with fiſh, attracts them in great
numbers ‖: in that of Panama, they are ſeen to
alight in bodies on the banks of pilchards left at
ſpring tides: And laſtly, all the ſhoals and adjacent iſlets are to ſuch a degree covered with
theſe birds, that their fat is melted for oil ¶.

The Pelican fiſhes in freſh water as well as in
the ſea. We need not, therefore, be ſurprized
to find it on the large rivers; but, what is ſingular, it does not confine itſelf to the contiguous, low, and wet grounds, but it frequents alſo
the drieſt countries, ſuch as Arabia and Perſia **,
where it is ſtiled *water-carrier* ††. As it is
obliged to place its neſt remote from the fountains
or wells where the caravans halt, it has been
obſerved to carry freſh water in its pouch from
a great diſtance to its young: and the good

---

* *Hiſt. Gen. des Voy. tom.* xiv. *p.* 456.
† *Note communicated by the Chevalier Deſhayes.*
‡ Dutertre.   § Labat.   ‖ Wafer.
¶ Oviedo.   ** Chardin.   †† *Tacab.*

Muſſulmans

Muffulmans fay, very pioufly, that God ordained
this bird to inhabit the defert, in order to provide drink for the parched pilgrims who journey
towards Mecca, as in ancient times he fent the
raven to feed Elias in the wildernefs *. Hence
the Egyptians, alluding to the manner in which
this large bird keeps the water in its bag, have
ftiled it the *river camel* †.

We muft not confound the *Barbary Pelican*,
mentioned by Dr. Shaw, with the real Pelican,
fince this traveller fays that it is not larger than
a lapwing. The Pelican of Kolben is only the
fpoon-bill. Pigafetta diftinguifhed well the
Pelican on the coaft of Angola, but was miftaken in beftowing that name on a bird of
Loango with tall legs like the heron ‡. We
doubt much alfo whether the *alcatraz*, which
fome travellers fay that they have feen in the
open fea between Africa and America §, be our
Pelican ; though the Spanifh inhabitants of the
Philippines and of Mexico have given it that
appellation : for the Pelican ftrays little from
the coafts, and when met with at fea it is regarded as a fign of the proximity of land ‖.

Of the two names *pelecanus* ¶ and *onocrotalus* **,

* Chardin.
† *Gemel el Bahr.* Vanfleb. *Voyage en Egypte, Paris,* 1677.
‡ *Hift. Gen. des Voy. tom.* iv. *p.* 588.
§ Id. *tom.* i. *p.* 448.
‖ Sloane.
¶ Ariftotle, *lib.* ix, 10.
** Pliny, *lib.* x. 47.

applied

applied by the ancients to this large bird, the latter refers to its ftrange voice, which they compared to the braying of an afs. Klein fuppofes that it makes this noife with its throat plunged in the water; but this idea feems to be borrowed from the bittern, for the Pelican utters its raucous cry far from the water, and fcreams loudeft in open air. Ælian defcribes and accurately characterizes the Pelican under the name of Κηλη *; but I cannot imagine why he reprefents it as an Indian bird, fince it is found now, and undoubtedly was found formerly, in Greece.

The firft name *pelecanus* or *pelicanus* has mifled the tranflators of Ariftotle, and even Cicero and Pliny: they have rendered it by the word *platea*, which would confound the *Pelican* with the *fpoon-bill*. When Ariftotle fays that the *pelecanus* fwallows thin fhell-fifh and cafts them up half-digefted, in order to feparate the meat which they contain, he imputes to it a habit which agrees better with the fpoon-bill, confidering the ftructure of its *æfophagus:* for the pouch of the Pelican is not a ftomach where digeftion is begun; and Pliny inaccurately compared the manner in which the *onocrotalus* fwallows and brings up its food to the procefs carried on in ruminating animals. " There is " nothing here," M. Perrault very judicioufly

* This word fignified any tumor, but more particularly a fwelling on the throat.

remarks, " but what enters into the general plan
" of the organization of birds : all of them have
" a craw in which their food is lodged ; in the
" Pelican it lies without and under the bill, in-
" ftead of being concealed within, and placed at
" the bottom of the *œfophagus*. But this exte-
" rior craw has not the digeftive heat of that of
" other birds, and in this bag the Pelican car-
" ries the fifh entire to its young. To difgorge
" them it preffes the pouch againft its breaft ;
" and this very natural act may have given rife
" to the fable fo generally told, that the Pelican
" opens its breaft to nourifh its offspring with
" its blood."

The neft of the Pelican is commonly found at
the verge of waters ; it places it flat on the
ground * ; and Salerne was miftaken, confound-

* Belon, Sonnerat, and others.—" They lay on the bare ground,
" and cover their eggs in this fituation. . . . I have found five
" under a female, which did not give herfelf the trouble to rife
" and let me pafs ; fhe only made fome pecks with her bill, and
" fcreamed when I ftruck her to drive her from her eggs. . . . There
" was a number of young ones on our iflet. . . . I took two young
" ones, which I faftened to a ftake, and I had the pleafure, for
" fome days, of feeing the mother, which fed them and remained
" the whole of the day with them, paffing the night on a tree
" above their heads ; all the three were grown fo familiar that they
" allowed me to touch them, and the young ones took very graci-
" oufly the little fifh which I offered them, and which they put firft
" into their pouch. I believe that I fhould have brought them
" away, if their dirtinefs had not hindered me : they are filthier than
" geefe or ducks ; and we may fay that their life is divided into
" three acts, to feek food, to fleep, and to eject every minute heaps
" of excrements as large as one's hand." Labat, *Nouveau Voyage
aux îles de l'Amerique*, tom. viii. *pp.* 294, 296.

ing

ing it probably with the spoon-bill, when he said that it breeds on trees. It is true that it perches on these, notwithstanding its weight and its broad webbed feet; and this habit, which would be less surprizing in those of America, where many aquatic birds perch, obtains equally in the Pelicans of Africa and of other parts of our continent *.

This bird, as voracious as it is destructive †, takes up in a single excursion as many fish as would feast half a dozen men. It swallows easily a fish of seven or eight pounds: and we are told that it also eats rats and other small animals ‡. Pison says, that he saw a kitten swallowed alive by a Pelican, which was so familiar that it walked into the market; where the fishermen hastened to tie its bag, left it should slily purloin some of their fish.

It eats with the side of its mouth, and when a person throws it a morsel, it snaps at it. The pouch in which it stores all its captures, consists

* "They are seen (in Guinea) to perch by the river side on "some tree, where they wait to shoot upon the fish which appear "on the surface." *Voyage de Gennes au Detroit de Magellan, Paris,* 1698, *p.* 41. "We saw those large birds called *pelicans* perch "upon trees, though they have feet like a goose. . . . Their eggs "are as large as a halfpenny roll." *Voyage à Madagascar, par Fr. Cauche, p.* 1361.

† *Inexplebile animal,* says Pliny.

‡ "It is exceedingly fond of rats, and swallows them entire. . . "sometimes we made it come near us, and as if it wished to amuse "us, it brought up a rat from its crop, and threw it at our feet." Bosman, *Voyage en Guinee, Lettre* xv.

of two ſkins; the inner coat is continued from
the membrane of the *œſophagus,* the outer is
only a production of the ſkin of the neck: the
wrinkles in which it is folded ſerve to contract
the bag, and when empty it becomes flaccid.
The bag of the Pelican is uſed as a tobacco-
pouch, and, in the French Weſt India iſlands is
termed *blague* or *blade* *, from the Engliſh word
bladder. It is aſſerted, that when theſe are
prepared, they are more beautiful and ſofter
than lamb-ſkins †. Some ſailors make caps of
them ‡; the Siameſe form muſical ſtrings of the
ſubſtance §; and the fiſhermen of the Nile uſe
the ſac attached to the jaw as a ſcoop for lading
their boats, or for holding water; as it neither
rots with moiſture nor can be penetrated by it ‖.

* *Blagues* are prepared by rubbing them well between the hands,
to ſoften the ſkin; and to increaſe the pliancy, they are beſmeared
with the butter of the cocoa, and again paſſed between the hands,
care being taken to preſerve the part which is covered with feathers
as an ornament. *Note communicated by the Chevalier Deſhayes.*—
" The ſailors kill the Pelican for its bag, into which they put
" a cannon-ball, and then hang it up, to give it the ſhape of a
" tobacco-pouch." Le Page du Pratz, *Hiſtoire de la Louiſiane, tom.*
ii. *p.* 113.

† " Our people killed many, not to eat ... but to have their
" *blagues*; for this is the name given to the pouch where they ſtore
" their fiſh. All our ſmokers uſe them to hold their cut tobacco.
" ... They are paſſed for lamb-ſkins, and they are much finer and
" ſofter; they become of the thickneſs of good parchment, but
" extremely pliant and ſoft. The Spaniſh women hem them very
" prettily and delicately with gold and ſilk; I have ſeen ſome
" pieces of work of this kind that were exceedingly beautiful."
Labat, *tom.* viii. *p.* 299.

‡ Cauche.　　§ Tachard.　　‖ Belon.

It

It feems that nature has provided with fingular caution againft the fuffocation of the Pelican: when, to fwallow its prey, it opens under water its whole bag, the *trachea arteria*, then leaving the *vertebræ* of the neck, adheres under this bag, and occafions a very fenfible fwelling; at the fame time two fphincter mufcles contract the *æfophagus* in fuch manner as to completely prevent the water from entering *. At the bottom of this fame bag is concealed a tongue fo fhort, that the bird has been believed to have none †; the noftrils alfo are almoft invifible, and placed at the root of the bill; the heart is very large; the kidney very fmall; the *cæca* equally fmall, and much lefs in proportion than in the goofe, the duck, and the fwan: Laftly, Aldrovandus affures us, that the Pelican has only twelve ribs; and he obferves that a ftrong membrane, furnifhed with thick mufcles, covers the pinions.

But a very interefting obfervation we owe to M. Mery and Father Tachard ‡, that air is

fpread

---

* Memoires de l'Academie des Sciences, *p.* 196.

† Gefner.

‡ " In a journey which we made to the loadftone-mine, M. de " la Marre wounded one of thefe large birds which our people call " *grand gofier*, and the Siameze *noktho* . . . its fpread wings mea- " fured feven feet and a half. . . . In diffection we found, under " the flefhy panicle, very delicate membranes which enveloped the " whole body, which folding differently, formed many confidera- " ble finufes, particularly between the thighs and the belly, be- " tween the wings and the ribs, and under the craw; fome were

" fo

spread under the skin through the whole body
of the Pelican. It may even be said that this is
a general fact, more conspicuous indeed in the
case of the Pelican, but which obtains in all
birds, and which M. Lory, a celebrated and
learned physician at Paris, has demonstrated by
tracing the communication from the atmosphere
to the bones and the pipes of the quills. In the
Pelican, the air passes from the breast into the
axillary sinuses, whence it insinuates into the
vesicles of the thick and swelled cellular membrane which covers the muscles and envelopes
the whole body, under the membrane in which
the feathers are rooted; these vesicles are inflated to such a degree, that on pressing the body, the air is observed to escape every way under
the fingers. During expiration, the air compressed in the breast passes into the sinuses, and thence
spreads into all the vesicles of the cellular texture : by blowing into the *trachea arteria*, we
may even make the course of the air sensible to
the eye. We may conceive therefore how

" so wide as to admit the two fingers ; these great sinuses divided
" into many little ducts, which by perpetual subdivision ran into
" an endless multitude of ramifications, which were perceptible only
" by the bubbles of air which inflated them ; insomuch that, pres-
" sing the body of this bird, one heard a little noise like that pro-
" duced by pressing the membranous parts of an animal which has
" been inflated. By the assistance of the probe and blowing, we
" discovered the communication of these membranes with the
" lungs." *Second Voyage of Father Tachard*; *Hist. Gen. des Voy.*
tom. ix. *p.* 311.

much

much the Pelican may enlarge its volume without increasing its weight, and how much this
must facilitate the flight of this great bird.

The flesh of the Pelican needed not to have
been forbidden among the Jews as unclean; for
it condemns itself by its bad taste, its marshy
smell, and its oily fat, though some navigators
have eaten of it*.

* " Their flesh is better than that of boobies or man of war
" birds."—*Dampier*.

[A] Specific character of the Pelican, *Pelecanus-Onocrotalus :*
" It is white, and its throat furnished with a pouch."

VARIETIES of the PELICAN.

WE have observed in many articles of this
Natural History, that in general the species of the large birds, like those of the large
quadrupeds, exist single, detached, and almost
without varieties; that they also appear every-
where the same; whereas under each genus or
in each family of small animals, and especially
in those of the little birds, there is a multitude
of breeds more or less akin to the parent stock,
and which have improperly been denominated
species. That term, and the metaphysical notion which it involves, often withdraws us from

the

the true knowledge of the fhades of nature in
her productions, much more than the names of
varieties, of *breeds*, and of *families*. But this
lineage, which is loft amidft the collateral
branches in the fmall fpecies, maintains itfelf
among the large ones; for they admit of few
varieties only, which may always be eafily referred to their primary trunk. The oftrich,
the caffowary, the condor, the fwan, all the
birds of the firft magnitude, have few or no varieties in their fpecies. Thofe which may be
reckoned the fecond order in bulk or ftrength,
fuch as the crane, the ftork, the pelican, the
albatrofs, admit of only a fmall number of thefe
varieties, which in the Pelican may be reduced
to two.

The BROWN PELICAN.

FIRST VARIETY

Pelecanus Fufcus. Linn. and Gmel.
Onocrotalus Fufcus. Briff.
The Pelican of America. Ellis and Edwards.
The Dufky Pelican. Pennant.

WE have already remarked, that the plumage
of the Pelican is fubject to vary, and that,
according to the age, it is more or lefs white

and

and tinged with a little rofe-colour : it feems to vary alfo from circumftances, for it is fometimes mixed with gray and black. Thefe differences have been remarked between individuals which undoubtedly belonged to the fame fpecies *. But thefe intermingled colours are fo little removed from a general gray or brown caft, that Klein has not hefitated to affert pofitively that the brown and white pelicans are only varieties of the fame fpecies. Sir Hans Sloane, who had carefully obferved the Brown Pelicans of America, confeffes alfo that they appeared to be the fame with the white pelicans. Oviedo, fpeaking of the pelicans with a cinereous plumage which occur on the rivers of the Antilles, remarks, that fome of them are of a very fine white †. We are inclined to think that the brown colour is the garb of the young ones ; for the Brown Pelicans have generally been found to be fmaller than the white. Thofe feen near Hudfon's Bay were alfo fmaller and of a dufky caft ‡ ; fo that their white is not occafioned by the feverity of the climate. The

---

* " Some had their plumage entirely white, with the light and " tranfparent caft of flefh-colour, except the wings, whofe great " quills had a tinge of gray and black ; the others were of a much " more decided flefh or rofe-colour." *Memoires de l'Academie des Sciences,* &c. The Pelican killed on the lake of Albufera had its back of a blackifh gray. *Journal Politique, &c.*

† Hift. Gen. des Voy. *tom.* xiii. *p.* 228.

‡ Ellis, and l'Hift. Gen. des Voy. *tom.* xiv. *p.* 663 ; and *tom.* xv. *p.* 268.

fame

fame variety of colour is obferved in the hot countries of the ancient continent. Sonnerat, after having defcribed two pelicans of the Philippine iflands, the one brown and the other rofe-colour, expreffes a fufpicion, as we do, that he had only viewed the fame bird at different ages. And what confirms our opinion, Briffon has given a Philippine pelican, which feems to form the intermediate fhade, being not entirely gray or brown, but having only the wings and part of the back of that colour, and the reft white *.

* " Above gray cinereous, below white, the rump of the fame " colour; the head and neck bright whitifh, with a longitudinal " bar on the upper part of the neck, variegated with brown and " whitifh; the greater wing-quills cinereous blackifh, the tail- " quills cinereous white, their fhafts blackifh, the lateral ones " bright white at their origin."

[A] Specific character of the *Pelecanus Fufcus:* " It is afh- " brown, its primary wing-quills black, its throat furnifhed with a " pouch."

The INDENTED-BILLED PELICAN.

SECOND VARIETY.

Pelecanus-Thagus. Gmel.
Onocrotalus Rostro Denticulato. Briff.
Onocrotalus Mexicanus Dentatus. Hernandez, &c.
The Saw-billed Pelican*. Lath.

IF the indenting of the bill of this Mexican Pelican be natural and regular, like that of the bill of the merganfer and fome other birds, this particular character would fuffice to conftitute a different fpecies, though Briffon gives it only as a variety: but if this indenting be formed by the accidental chipping of the edges of the bill, as we have remarked in the bill of certain calaos, the accidental difference deferves not even to be admitted as a variety; and we lean more to this opinion, as Hernandez mentions the common pelican and the Indented-billed Pelican as inhabiting the fame places.

---

* *Atototl, Alcatraz, Onocrotalus Mexicanus Dentatus.* Hernandez.— *Atototl.* Fernandez.

The CORMORANT.

Le Cormoran. *Buff.*

Pelecanus-Carbo. Linn. and Gmel.
Phalacrocorax. Gefner and Briffon.
Morfex. Gefn. and Aldrov.
Carbo Aquaticus. Gefner.
Corvus Aquaticus. Gefner. Aldrov. Johnft. &c. *

THE name of this bird was· formerly pronounced *cormaran* or *cormarin*, being contracted from *corvus marinus* or *fea-raven* † : the
Greeks ftiled it the *bald-raven*. Yet it refembles the raven in nothing but its black plumage,
and even this is downy, and of a lighter fhade.

The Cormorant is a pretty large bird with

---

* In Greek Φαλακροκοραξ, from φαλακρος *bald*, and κοραξ *a
raven*. The Spanifh name *Cuervo Calvo* has the fame fignification : and the notion of *water* or *fea-raven* is implied in modern
Latin, in Italian, in German, and Silefian, by the appellations of
Corvus Aquaticus, Corvo Marino, Waffer-Rabe, See-Rabe. In
Swedifh it is termed *Hafts-tjaeder:* in Norwegian *Skary*; and in
the ifle of Feroe *Hupling:* in Poiifh *Krukwodny.* In fome of the
French provinces it is ftiled *Crot-Pefcherot*, or *Dirt-fifher.*

† *Caius*, or Dr. Kay, in Gefner, conjectures, that *cormorant* is a
corruption of *corvorant, corvus vorans*, or *devouring raven:* and
Pennant and Latham have adopted *corvorant.* But it is doubtful
whether, for the fake of a fpecious etymology, we fhould alter a
word of fuch common ufe in our language : the derivation affigned
in the text is befides more probable.—*T.*

webbed

THE CORMORANT.

webbed feet, which fwims and dives with equal facility, and devours multitudes of fifh: it is nearly of the bulk of the goofe, of a narrower form, rather thin than thick, and lengthened by a large tail more fpread than ufual in the aquatic birds; this tail confifts of fourteen ftiff feathers like thofe of the woodpecker's tail; they are black gloffed with green, like almoft all the reft of the plumage: the back is waved with black feftoons on a brown ground; but thefe fhades vary in different individuals, for Salerne fays, that the colour of the plumage is fome-times a greenifh black: all of them have two white fpots on the outfide of the legs; with a white gorget, which embraces the top of the neck like a chin-piece: there are white feathery filaments like briftles, ftuck on the top of the neck and the upper part of the head, of which the front and the fides are bald *: a fkin, alfo naked, clothes the under fide of the bill, which is ftraight to the point, where it is bent into a very fharp hook.

The Cormorant is one of the few birds which have four toes connected together by a fingle piece of membrane. We might thence infer that it is a very great fwimmer; yet it remains lefs in the water than many other aquatic birds,

---

* " Some animals are naturally bald, as the oftriches and the " water ravens, which thence derive their Greek name." *Plin.* lib. ii. 38.

whofe

whofe foles are neither fo continuous nor fo broad: it frequently flies and perches on trees. Ariftotle afcribes this habit to it alone of all the *palmiped* birds *; but it is common to the pelican, the booby, the frigate, the anhinga, and the tropic bird; and what is fingular, thefe birds, together with it, form the fmall number of the aquatic fpecies which have the four toes connected by continuous membranes. This coincidence has induced modern ornithologifts to range five or fix birds under the generic name of *pelican* †. But the analogy muft be ftrained for the fake of a fcholaftic generalization; when, from the refemblance of a fingle part, the fame-appellation is given to fpecies fo different from each other as that of the tropic bird, for inftance, and that of the common pelican.

The Cormorant is fo dextrous in fifhing, and fo voracious, that when it vifits a pool, it commits alone more havoc than a whole flock of other pifcivorous birds. Fortunately, it refides almoft conftantly on the fea-fhores, and feldom occurs in inland countries ‡. As it can remain a long time plunged §, and fwims under water

---

* *Hift. Animal.* lib. viii. 3.

† Klein and Linnæus have formed this family. The Cormorant there figures, under the appellation *pelecanus carbo*; the frigate, under that of *pelecanus aquilus*, &c.

‡ " On the 27th of January (1779) a Cormorant was brought " to me, which was juft killed on the fide of the river Ouche: it " had perched on a willow." *Extract of a letter from M. Hebert.*

§ Schwenckfeld.

with

with the rapidity of a dart, its prey ſcarce ever eſcapes, and it almoſt always emerges holding a fiſh acroſs in its bill : to ſwallow the victim it employs a ſingular expedient; it toſſes up the fiſh in the air, and dextrouſly catches the head in falling, ſo that the fins lie flat and favour the paſſage down the throat, while the membranous ſkin that lines the under ſide of its bill ſtretches to admit the whole body of the fiſh, which is often very large in proportion to the neck of the bird.

In ſome countries, as in China, and formerly in England *, the ſkill of the Cormorant in fiſhing was turned to profit : for, by buckling a ring about the lower part of its neck, to prevent deglutition, and accuſtoming it to return with its acquiſitions in its bill to its maſter, it was made, ſo to ſay, a domeſtic fiſher. On the rivers of China, the Cormorants, thus buckled, are perched on the prows of the boats, and on a ſignal being given, by ſtriking the water with an oar, they plunge into that element, and quickly emerge with a fiſh, which is taken out of their bill; and this toil is continued, till its maſter, ſatisfied with the earnings, looſens its collar, and permits it to fiſh for its own account †.

Hunger alone gives activity to the Cormorant; it becomes lazy and ſluggiſh after its appetite is glutted. It inclines to fat, and though

* According to Lynceus in Willughby.
† Nieremberg. Voyage a la Chine, par de Feynes: Paris, 1630, p. 173. Hiſt. Gen. des Voy. tom. vi. p. 221.

3　　　　　　　　　　　　　it

it has a very ſtrong ſmell, and an unpleaſant
taſte, it is not always deſpiſed by ſailors, to whom
the ſimpleſt and coarſeſt fare is often more delicious than the moſt exquiſite viands to our delicate palates *.

The Cormorant occurs in the remoteſt latitudes; in the Philippines †, in New Holland ‡,
and even in New Zealand §.   In the bay of Saldana there is an iſland ſtiled the *Iſland of Cormo-*
rants ‖, becauſe it is covered, as it were, with
theſe birds.   They are not leſs common in other
parts near the Cape of Good Hope.   " Some-
" times flocks," ſays the Viſcount de Querhoënt,
" are ſeen of two or three hundred in the road
" off the Cape.   They are not timid; which
" is no doubt becauſe they are little moleſted.
" They are naturally indolent: I have ſeen
" them reſt ſix hours on the buoys of our an-
" chors.   Their bill is furniſhed below with a
" ſkin of a fine orange colour, which extends
" under the throat a few lines, and dilates at
" pleaſure.   The iris is of a fine light green;

* " Their fleſh has an exceeding rank fiſhy taſte; however it is
" pretty good, being very fat." *Dampier.*—" We killed a great
" number of Cormorants, which we ſaw perched on their neſts in
" the trees, and which were roaſted or dreſſed in a ſtove, and af-
" forded us excellent diſhes." *Cook's Firſt Voyage.*

† It is there called *Colocolo.* See Philoſ. Tranſ. N° 285, and
Hiſt. Gen. des Voy. *tom. x. p.* 412.

‡ Cook.

§ Ibid.

* Flacourt.

" the

" the pupil black; the orbit edged with a violet
" fkin: the tail is formed like that of the wood-
" pecker, containing fourteen hard fharp quills.
" The old ones are entirely black; but the
" young ones are all gray the firft year, and
" have not the orange fkin under the bill.—
" They were all very fat *."

The Cormorants are alfo very numerous in
Senegal, according to M. Adanfon †. They
feem alfo to be the *Plutons* of the ifland Mauritius, as defcribed by the traveller Leguat ‡; and,

---

* Remarks made in 1774, by the Vifcount de Querhoent, of his
majefty's navy.

† " We arrived on the 8th October at Lamnai (a little ifland of
" the Niger); the trees were there covered with fuch a prodigious
" multitude of Cormorants, that the Laptots, in lefs than half an
" hour, filled a canoe with young ones which had been taken by the
" hand, or felled with fticks, and with old ones, of which feveral
" dozens fell at every fhot." *Voyage au Senegal, p.* 80.

‡ " On a rock, near the ifland Mauritius, came birds which we
" called *Plutons* (Pluto's), becaufe they are entirely black like *ra-*
" *vens*; they have alfo nearly the fame fhape and fize, but their
" bill is longer, and hooked at the end; their foot is that of the duck:
" thefe birds live fix months of the year at fea; thofe in the neigh-
" bourhood repaired to our rock, where they hatched. They have
" a cry as ftrong almoft as the lowing of a calf, and they make a great
" noife in the night. During the day they were very ftill, and fo
" tame that they fuffered us to take the eggs from under them with-
" out ftirring; they lay in holes of the moft projecting rock that
" they can find. Thefe birds are very fat, and ill-tafted, unwhole-
" fome, and abominably ftinking. Though their eggs are hardly
" better than their flefh, we ate them in neceffity; they are white,
" and as large as thofe of our *hens*; when thefe were taken, they
" retired into their holes, and fought with each other, till they
" were all over bloody." *Voyage de François Leguat*; *Amfterdam,*
1708, *tom.* ii. *pp.* 45, 46.

what

what is fingular, they fupport alike the heat of that climate and the cold of Siberia. It appears, however, that the fevere winters of the northern regions oblige them to migrate: for thofe which in fummer inhabit the lakes in the neighbourhood of Selenginfkoi, where they are called *baclans*, have been obferved to retire in autumn to Lake Baikal *, there to fpend the winter. The fame muft be the cafe with the *ouriles*, or Cormorants of Kamtfchatka, well defcribed by Krafcheninicoff †, and indicated in the fabulous relation of the Kamtfchadales, who fay that thefe birds have bartered their tongue with the wild goats, for the tufts of white briftles on their neck and thighs ‡: yet is it falfe that thefe birds have no tongue; and Steller avers, that they found day and night with a voice like the note of a little hoarfe trumpet.

Thefe Cormorants of Kamtfchatka pafs the night, gathered in flocks, on the projections of craggy rocks, from which they often fall to the ground during their fleep, and then become the prey of the foxes, which are ever on the watch. In the day-time, the Kamtfchadales fearch for their

* " The inhabitants of thefe cantons believe that when the *bac-* " *lans* make their nefts on the top of a tree, it grows dry; in fact, " we faw that all the trees where were nefts of thefe birds had wi- " thered; but perhaps they chufe trees already decayed." Gmelin; *Voyage en Siberie, tom.* i. *p.* 244.

† Fllfa Gen. des Voy. *tom.* xix. *p.* 272.

‡ Idem. *tom.* i. *p.* 272.

eggs, at the rifk of tumbling upon the precipices, or dafhing into the fea. And to catch the birds themfelves, they faften a running knot to the end of a rod: the heavy, indolent Cormorant, when once feated, cares not to ftir, but only turns his head from right to left to avoid the noofe, which is at laft flipped on his neck.

The head of the Cormorant is fenfibly flat, like that of moft diving birds; its eyes are placed very much forward, and near the corners of the bill, whofe fubftance is very hard, and fhining like horn; the feet are black, fhort, and very ftrong; the tarfus is very broad, and flattened fidewife; the middle nail is ferrated interiorly, like that of the heron; the pinions are very long, but clothed with fhort quills, which makes it fly heavily, as Schwenckfeld obferves. This natu-ralift is the only perfon who afferts that he faw a particular little bone, which rifing behind the *cranium*, defcends in form of a thin blade, and is inferted into the mufcles of the neck.

[A] Specific character of the Cormorant, *Pelecanus Carbo :* " Its " tail is rounded, its body black, its head fomewhat crefted." It makes its neft with fticks, fea-weed, grafs, &c. and lays fix or feven eggs, which are white and of an oblong fhape.

The S H A G.

LE PETIT CORMORAN, *ou* LE NIGAUD *.

Buff.

Pelecanus Graculus. Linn. and Gmel.
Phalacrocorax Minor. Briss.
Graculus Palmipes. } Aldrov. Sibb. &c.
Corvus Aquaticus Minor.
The Shag, or Crane. Ray and Will.

THE heaviness or rather indolence natural to
all the Cormorants, is still more remarkable in the present; which has, for that reason,
been styled the *Shag* or *Ninny (Niais* ou *Nigaud)*.
This species is not less diffused than the former: it occurs particularly in the islands and the
extremities of the southern continents. Cook
and Forster found it on the island of Georgia;
which, though not inhabited, and almost inacces-
sible by man, is stocked with these little Cormorants, which share the domain with the penguins,
and lodge among the tufts of rushy grass, the
only vegetable production in that dreary tract.
Staten-land is similar, and contains likewise great
numbers of birds. An island in the Straits of
Magellan was so full of them, that Captain Cook
called it *Shag Island*. It is in these extremities

* *i. e.* The Little Cormorant, or Simpleton.

of

of the globe that nature, benumbed with cold,
has allowed five or fix fpecies ftill to fubfift, the
laft inhabitants of the territories invaded by the
progrefs of refrigeration : they live in calm apa-
thy, the gloomy prelude of the eternal filence
which foon will there eftablifh its iron reign *.

" One

---

* The gradual refrigeration of the earth is a favourite hypothefis
of our ingenious author. He had fuppofed that a comet, reeling in
its eccentric orbit, dafhed againft the fun, and ftruck off that ignit-
ed matter, which, gathering into globes and recovering from diforder, formed our planetary fyftem. He made a great many experiments with heated metallic balls of different diameters, to difcover their rate of cooling : and in his Epochs of Nature, the moft
fanciful of all his works, he transferred thefe deductions to the globe
of the earth ; he poetically delineated its condition at the various
ftages of cooling ; and he predicted the glacial cruft, which, in procefs of ages, will imprifon old ocean.

But this account of the formation of the world is totally incon-
fiftent with the eftablifhed laws of motion ; and even were it admitted, it would only remove the difficulty a fingle ftep ; for how
was the comet produced ? The experiments with heated balls are
inconclufive. Bodies have no natural tendency to cool, any more
than to heat : they only maintain an equality of temperature with
the furrounding matter. In ordinary cafes, cooling is produced
by the fucceffive application of different portions of air to the hot
furface. A body colder than the atmofphere would in the fame
manner be heated. In the exhaufted receiver of an air-pump, the
progrefs of heating or cooling is much flower ; and could a perfect
vacuum be obtained, there is every reafon to think that a body would
for ever retain the fame temperature. The earth would therefore
preferve perpetually its heat ; and even though we fhould fuppofe it
to be environed with a fubtle æther (which is altogether improbable) the communication of this heat to other planets or fyftems
would be extremely flow and imperceptible. Nay, if there be any
difference, the earth is growing warmer, by the inceffant abforption
of the fun's rays: yet fo vaft is its mafs, that this effect will not much
exceed a degree in a thoufand years. Hiftory feems to corroborate this

U 2   conjecture.

" One is aftonifhed," fays Captain Cook, " at
" the peace which prevails in this land. The
" animals that inhabit it would feem to have
" formed a league not to difturb their mutual
" tranquillity. The fea-lions occupy the greateft
" part of the coaft; the white bears refide in
" the interior part of the ifland; and the Shags
" lodge in the loftieft rocks : the penguins fettle
" where they have eafieft communication with
" the fea; and the other birds chufe places more
" retired. We have feen all thefe animals in-
" termingled and walking together, like cattle
" or poultry in a farm-yard, without offering
" the leaft injury to each other."

In thefe dreary waftes, naked, and almoft
frozen, the Shags breed in the ragged fides of
rocks, or the projecting cliffs that overhang the
ocean. In fome parts, their nefts are found
among fmall patches of flags, or in the tall tufts
of the coarfe grafs which we have mentioned.
There they inhabit, collected in thoufands: the
report of a mufket does not difperfe them; they
only rife a few feet, and alight again into their
nefts. Nor need we ufe fire-arms, for they may

conjecture. Witnefs the ancient and modern ftate of Italy, a country which has remained nearly in the fame ftate of cultivation.

With regard to the huge bones dug up in Siberia, and believed to
be thofe of the elephant, which is a native of hot climates, they probably belong to fome animal whofe fpecies is now extinct. Such at
leaft was the decifion of the celebrated Dr. Hunter, upon examining
the bones found near the Ohio, in the back parts of North America. T.

be felled with sticks, and yet their companions will not be alarmed, or endeavour to escape from the massacre. Their flesh, especially that of the young ones, is pretty good food.

These birds do not stray far into the sea, and seldom lose sight of land. Like the penguins, they are clothed with a very thick plumage, well adapted to guard against the severe and continual cold of the frozen regions which they inhabit. Forster seems to admit several species or varieties of this bird; but as he does not sufficiently distinguish them, and as the different mode of nestling on tufts or in the crevices of rocks is insufficient to discriminate the species, we shall describe only the common Shag known in our climates.

They are pretty numerous on the coast of Cornwal and in the Irish sea, particularly on the Isle of Man *. They are found also on the shores of Prussia †, and in Holland ‡ near Sevenhuis, where they breed on tall trees. Willughby says, that they swim with their body entirely immersed, and only their head out of the water; and that they are as nimble and alert in that element as they are sluggish on land, and escape the shot by diving the instant they perceive the flash. In general, the Shag has the same natural habits with those of the cormo-

* Ray.　　† Klein.　　‡ Ray.

U 3　　　　　　rant,

rant *, which it refembles in its figure and in
its colours: the difference confifts in this, that
its body and limbs are fmaller and more flender,
its plumage brown under the body, its throat
not naked, and that there are only twelve quills
in the tail †.

Some ornithologifts have ftiled the Shag the
palmiped jay: but this is as little proper as the
vulgar appellation of *water raven* given to the
cormorant. The *palmiped jays* which Captain
Wallis met with in the Pacific Ocean ‡ are
probably a fpecies of Shag; and to it we fhall
alfo refer the *handfome cormorants* of which
Captain Cook faw large flocks neftled in fmall
cavities, which thefe birds feemed to have wi-
dened for themfelves in a rock of fchift, whofe
broken fides terminate New Zealand.

The interior organization of the Shag prefents many curious particulars, which we fhall
extract from the obfervations of the Academicians. A bony ring embraces the *trachea arteria* above the bifurcation: the *pylorus* is not
inferted at the bottom of the ftomach as ufual,
but opens into the middle of that ventricle,
leaving one - half hanging below; and this
lower part is very flefhy and mufcular, fo as to

* " To fwallow a fifh it toffes it into the air, and catches with
" its bill the head foremoft. We have feen it perform this ma-
" nœuvre with fuch addrefs, that it never miffed." *Anciens Mé-
moires de l'Academie des Sciences, tom.* iii. *partie* 3. *p.* 214.
† Ray and Willughby.
‡ In latitude 20 degrees 50 minutes north.

force up by its contraction the food to the orifice of the *pylorus:* if we blow into the *œsophagus*, it fwells and appears a continuation of the ftomach, which otherwife is feparated from it by a narrow ring: the inteftines are inclofed in an *epiploon*, well lined with fat, of the confiftence of tallow; this fact is an exception to what Pliny fays in general of oviparous animals, that they have no epiploon *. The fhape of the kidnies is fingular; they are not parted into three lobes, as in other birds, but jagged like a cock's comb on their convex portion, and divided from the reft of the lower belly by a membrane which invefts them: the *cornea* of the eye is of a bright red, and the cryftalline approaches the fpherical form, as in fifh: the bafe of the bill is furnifhed with a red fkin, which alfo furrounds the eye: the aperture of the noftrils is fo narrow a flit as to have efcaped obfervers, who have afferted, that the cormorants, both the greater and the leffer, want the noftrils: the greateft toe in thefe two fpecies is the outer, compofed of five *phalanges*, the next one containing three, the third three, and the laft, which is the fhorteft, only two: the feet are of a fhining black, and armed with pointed nails †:

* Lib. ii. 37.

† M. Perrault refutes ferioufly the fable of Gefner, who fays, that there is a kind of cormorant which has a membranous foot, with which it fwims, and another whofe toes are naked, with which it feizes its prey.

under

under the feathers there is a very fine down, as thick as that of the fwan; fmall filky feathers, clofe like velvet, cover the head; from which Perrault infers, that the cormorant is not the bald raven, *phalacrocorax*, of the ancients. But he·ought to have qualified this affertion, having himfelf obferved before, that on the fea-fhores there occurs a great cormorant different from the fmall cormorant or fhag : and this bald-headed great cormorant is, as we have feen, the true *phalacrocorax* of the ancients.

[A] Specific chara&er of the Shag, *Pelecanus-Graculus :* " Its " tail is rounded; its body black, below brown; it has twelve " wing-quills."

The SEA SWALLOWS.

LE HIRONDELLES DE MER. *Buff.*

OF the multitude of names * transferred for
the moſt part improperly from the land
animals to thoſe of the ſea, a few have been
happily applied; ſuch as that of the Swallow,
given to a ſmall family of piſcivorous birds,
which reſemble our ſwallows by their long
wings and forked tail, and by their continual
circling on the ſurface of the water. As the
land ſwallows flutter ſwiftly in the fields or
round our dwellings in ſearch of winged inſects,
ſo the Sea Swallows circle and glance rapidly on
the liquid plains, and nimbly ſnatch the little
fiſh which play on the ſurface. Such reſemblance in the form and habits of theſe two
kinds of birds might, in ſome meaſure, juſtify
their receiving the ſame appellation: yet they
differ eſſentially in the ſhape of their bill, and
the ſtructure of their feet. The Sea Swallows
have ſmall membranes ſhrunk between their

* In German *See Schwalbe:* in Swediſh, and the other northern
languages, *Taern, Terns, Stirn;* whence Turner derives the name
Sterna, adopted by nomenclators to diſtinguiſh this genus of
birds.

toes,

toes, which are not adapted for swimming * :
for nature seems to have bestowed on these
birds only the power of their wings, which are
extremely long, and scooped like those of the
common swallows. They likewise glide and
circle, sink and rise in the air, crossing and en-
twining their various irregular track in a thousand directions † ; their flight is impelled by
starts of momentary caprice, and led by the
sudden glimpse of their fugitive prey. They
snatch the victim on wing, or alight only a
moment on the surface ; for they are averse to
swim, though their half-webbed feet might contribute to that purpose. They reside commonly
on the sea-shores, and frequent also lakes and
great rivers. The Sea Swallows, in flying,
scream loud and shrill like the martins, especially when, in calm weather, they rise to a great
height in the air, or when they congregate in
summer to make distant excursions, but particularly in the breeding season, at which time
they are more than ever restless and clamorous,
perpetually redoubling their motions and their
cries : and as they are always extremely numerous, we can hardly, without being stunned with

---

* Hence Aldrovandus, looking upon the Sea Swallows as little
gulls, distinguishes them by the name of cloven-footed gulls, *lib.*
xix. 10. *de laris fissipedibus.*

† " Sailors call those nimble birds found at sea, *croiseurs (crossers)*
" when they are large ; *goelettes* when they are small." *Remarks
made by the Viscount de Querhoënt.*

the

the noife, approach the fhore, where they have
difpofed their eggs or collected their young.
They arrive in flocks on our weftern coafts in
the beginning of May *: moft of them remain
without quitting the beach; others advance
farther, and following the rivers, feek the lakes
and the large pools †. Every where they live
on fmall fifh, and fometimes they even fnap
winged infects in the air. The report of fire-
arms does not intimidate them, and this fignal
of danger, fo far from driving them off, feems
rather to attract them; for the inftant the
fowler hits one of a flock, the reft croud about
their wounded companion, and drop with it to
the furface of the water. The land fwallows
are likewife remarked to gather at the noife of
a gun, or at leaft they are not frighted away.
Does not this habit proceed from a blind fecurity? Birds that are hurried inceffantly with a
rapid flight, are more incautious than fuch as
fquat in the furrows or perch on the trees; they
have not learnt, like thefe, to obferve and diftinguifh us, and to fly from their moft dangerous enemies.

The feet of the Sea Swallows differ not from
thofe of the land fwallows, except that they are
femi-palmated; for in both they are very fmall

* Obfervation made on thofe of Picardy, by M. Baillon.

† As that of Indre, near Dieuze in Lorraine, which, including
its windings and inlets, is feven leagues in compafs.

and

and fhort, and unfit for walking. The pointed
nails which arm the toes feem not more neceffary to the Sea Swallow than to the land one,
fince thefe birds equally feize the prey with the
bill : that of the Sea Swallow is ftraight, tapered to a point, fmooth, not indented, and flat
at the fides. The wings are fo long, that the
bird when at reft feems incumbered by them,
and in the air appears all wing. But if this
great power of flight makes the Sea Swallow an
inhabitant of the air, it has other properties that
diftinguifh it as an inhabitant of the water : for,
befides the fcalloped membrane between the
toes, a fmall portion of the leg, as in almoft all
the aquatic birds, is bare, and the body is covered with a thick and very clofe down.

This family of Sea Swallows includes a number of fpecies, moft of which have croffed the
ocean, and ftocked its fhores. They are found
fpread from the feas, the lakes *, and the great
rivers † of the north, as far as the vaft bounda-
ries of the Southern Ocean ‡ ; and they occur

in

* Even their name *taern* or *terns*, in the northern languages, fignifies *lake*.

† Gmelin fays, that he faw innumerable flocks of them on the
Jenifea, near Mangafea in Siberia. *Voyage en Siberie*, tom. ii. *p.* 56.

‡ Captain Cook faw Sea Swallows near the Marquefas, which
are iflands feen by Mendana.—The fame navigator was attended
by thefe birds from the Cape of Good Hope to the 41ft degree of
fouth latitude.—Captain Wallis met with them in the 27th degree
of latitude and the 106th degree of weft longitude, on the great
South Sea.—" The low iflands within the tropics, and the whole of
" the

in almoſt all the intermediate regions *. We
ſhall adduce the proofs in deſcribing the different ſpecies.

" the Archipelago which ſurrounds Otaheite, are filled with
" flights of Sea Swallows, boobies, frigats, &c." *Forſter.*—
" The Sea Swallows rooſt under the buſhes in Otaheite; Mr.
" Forſter, in an excurſion before ſun-riſe, took ſeveral that were
" ſleeping along the road." *Cook.*

* Sea Swallows are found in the Philippines, in Guiana, and
Aſcenſion. We may recognize them in Dampier's deſcription of
birds which he met with near New Guinea. " On the 30th of July,
" all the birds which had hitherto accompanied us, quitted the
" veſſel; but we ſaw others of a different kind, which were as large
" as lapwings, with a gray plumage, the ſpace about their eyes
" black, their bill red and pointed, their wings long, and their tail
" forked as in ſwallows."—" On the 13th of July 1773, in lati-
" tude 35° 02′, and longitude 2° 48′, during a violent north-weſt
" wind, M. de Querhoënt ſaw many petrels and Sea Swallows; theſe
" were at leaſt a half ſmaller than the petrels; their wings were
" very long, and ſhaped like thoſe of our martin: they uſually keep
" in flocks, and come very near veſſels."

The GREAT SEA SWALLOW.

Le Pierre-Garin, *ou* La Grande Hiron-delle de Mer de nos Côtes. *Buff.*

FIRST SPECIES.

Sterna-Hirundo. Linn. and Gmel.
Sterna Major. Briff.
Sterna. Gefner, Aldrov. Johnft. &c.
Larus Hirundo. Kramer.
Larus Albicans. Marfigli, and Klein.
Hirundo Marina Major. Will. and Sibb.
The Greater Tern. Pennant.
The Common Tern *. Latham.

THIS is the largeft fpecies of Sea Swallows that appears on our coafts, being near thirteen inches from the end of the bill to that of the nails, near fixteen to the end of the tail, and almoft two feet acrofs the wings: its flender ftature, its handfome gray mantle, its black cap, and its red bill and feet, confpire to make it a beautiful bird.

On the return of fpring, thefe Swallows arrive

---

* In Swedifh *Tœrna:* in Norwegian *Terne, Tende, Tendelobe, Sand-tolle, Sand Tœrna:* in Danifh *Tœrna:* in Dutch *Iflerre:* in Swifs *Schirring:* in Polifh *Jaſkolla-morſka,* or *Kulig-morſki:* in Icelandic *Therne, Krua:* in Lapponic *Zhierrek:* in Greenlandic *Emerkotulak.*

in

THE COMMON TERN.

in great flocks on our maritime shores, where they separate into troops: some penetrate into the interior provinces, such as the Orleanois *, Lorraine †, Alsace ‡, and perhaps farther, following the course of the rivers, and settling on the lakes and great pools ; but the greater part remain on the coasts, and make long excursions into the sea. Ray observes, that they are usually found fifty leagues from the most western part of England, and are even met with the whole way to Madeira ; and that a vast multitude resort to breed on the *Salvages*, desert islets at a small distance from the Canaries.

On the coasts of Picardy, these birds are named *pierre garins*; they are lively and agile, says M. Baillon, venturous and skilful fishers: they dart after their prey into the sea, emerge again in an instant, and mount to their former height in the air. They digest the fish as quickly almost as they catch it ; the part which touches the bottom of the stomach dissolves first: the same effect is observed in herons and gulls. So great is the digestive power, that the Sea Swallow can, after the interval of an hour or two, make a second meal. They fight frequently, quarrelling about their prey. They swallow fish more than an inch thick, and so

* Salerne.
† Lottinger.
‡ On the Rhine near Strasburg, where they are called *Speurer*, according to Gesner.

long,

long, that the tail projects out of their bill.
Those that are taken and sometimes fed in gardens *, refuse not flesh, which they will not
touch in the state of liberty.

These birds pair on their arrival about the
first days of May. Each female drops in a small
hole on the naked sand two or three eggs, very
large in proportion to her bulk. The place
chosen by them for this purpose is always screen-
ed from the north wind, and situated below some
downs. If a person approach the nests, the parents will rush precipitately from aloft, and flutter round him with loud reiterated screams of
anger and inquietude.

Their eggs are not all of the same colour,
some being very brown, others gray, and others
almost greenish : these last probably belong to
young pairs; for they are rather smaller, and it is
known that of all birds which have coloured
eggs, those of old ones are deeper stained, rather
thicker, and less pointed than those of young
ones, especially in their first layings. The female of this species covers only during the
night, or in the day when it rains : at all other

* " I have had several in my garden, where I could not keep them
" long, because of the annoyance of their perpetual cries, which were
" continued even during the night. These captive birds lost almost
" entirely their chearfulness ; formed to sport in the air, they feel
" incumbered on the ground, their short feet hamper them on every
" obstacle they meet." *Extract of a Memoir of M. Baillon, on the
Common Terns.* from which we take the details of the history of
these birds.

times

times fhe leaves her eggs to the heat of the
fun. " When the fpring is fine," M. Baillon
writes me, " and the incubation was begun in
" warm weather, the three eggs, their ufual
" number, are hatched in three fucceffive
" days, in the order they were laid; the deve-
" lopement in the two firft being forwarded by
" the influence of the folar beams. If the wea-
" ther was rainy or cloudy in the commencement,
" that effect is not perceived, and the eggs burft
" together. The fame remark has been made
" with regard to fea-larks and fea-pies; and it
" may reafonably be extended to all birds which
" lay on the naked beach.

" The young Sea Swallows, when juft hatch-
" ed, are clothed with a thick down, light gray,
" and fprinkled with fome black fpots on the
" head and the back. Their parents fetch
" them bits of fifh, particularly liver and gills:
" and when the mother comes at night to co-
" ver the unhatched egg, the callow chicks
" creep under her wings. Thefe maternal cares
" laft but a few days; the young affemble at
" night, and lie clofe together. Nor do the
" parents long nourifh them by the bill: with-
" out defcending each time to the ground, they
" drop, or fo to fpeak, rain upon them food: the
" young ones, now voracious, fight and quarrel
" with each other, and fcream loudly. Yet the
" parents continue to watch them from aloft in
" the air; a cry which they give as they glide

VOL. VIII. x " along

" along conveys the alarm, and inftantly the
" brood fquat clofe on the fand. It would be
" difficult to difcover them, did not the fhrieks
" of the mother betray the fpots where they
" lurk. They make no effort to efcape, but
" may be gathered by the hand like ftones.

" They fly not till more than fix weeks after
" they are hatched, it requiring all that time for
" their broad wings to grow; like the land
" fwallows, which remain longer in the neft
" than other birds of the fame fize, and fally
" out better feathered. The firft feathers of
" the young Terns are light gray on the head, the
" back, and the wings; the true colours appear
" not till after moulting. But they have all
" the fame colours when they return in fpring.
" They depart from the coafts of Picardy about
" the middle of Auguft; and I remarked that laft
" year, 1779, they chofe a north-eaft wind."

[A] Specific chara&ter of the Common Tern, *Sterna-Hirundo*:
" Its two outermoft tail-quills are parted with black and white."

The LESSER SEA SWALLOW.

LA PETITE HIRONDELLE DE MER. *Buff*.

SECOND SPECIES.

Sterna Minuta. Linn. and Gmel.
Sterna Minor. Briff.
Larus Pifcator. Gefner, Aldrov. Johnft. &c.
The Leffer Tern *. Stillingfleet, Penn. and Lath.

THIS little Sea Swallow refembles the preceding fo clofely in its colours, that they are diftinguifhed only by their conftant and confiderable difference of fize. The prefent is not larger than a lark, though as clamorous and roving as the firft fpecies †. Yet will it live a prifoner, if caught in a fnare. In Belon's time, the fifhermen floated a crofs of wood, in the middle of which was faftened a fmall fifh for bait, with limed twigs ftuck to the four corners, on which the bird darting was entangled by the wings. Thefe little Sea Swallows, as well as the great ones, frequent our feas, lakes, and rivers, and retire alfo on the approach of winter.

* Near Strafburg it is called *Fifcherlin :* in Polifh *Ribtu.*
† " It is fo noify as to ftun the air, and to moleft the people who " pafs the fummer near marfhes and brooks." *Belon.*

[A] Specific character of the Leffer Tern, *Sterna Minuta :* " Its " body is white, its back hoary, its front and eye-brows white."

The GUIFETTE.

THIRD SPECIES.

Sterna Nævia. Gmel. and Briff.
Rallus Lariformis. Linn.
The Kirr-Meuw. Klein.
The Cloven-footed Gull. Albin.
The Kamtfchatkan Tern. Penn. and Lath.

THIS Sea Swallow is named *Guifette* on the
coaft of Picardy. Its plumage, which is
white under the body, is agreeably variegated
with black behind the head, with brown clouded
with rufty on the back, and with a handfome
gray fringed with whitifh on the wings. It is
of a middle fize between the two preceding, but
differs in feveral particulars with regard to habits and œconomy. Baillon, who compares it
with the great fea-fwallow, fays, that it is diftinguifhed by many charaĉters: 1. It does not
continually feek its food on the fea; it is not pifcivorous, but rather infeĉtivorous, feeding as
much on flies and other infeĉts, which it fnaps
in the air, as on thofe which it catches on the
water: 2. It is not fo clamorous as the great
fea-fwallow: 3. It does not lay on the naked
fand, but chufes in the marfhes a tuft of herbs

*

or

or mofs in fome infulated hillock amidft the water or on its brink; it carries thither fome dry ftalks of herbs, and drops its eggs, which are generally three in number: 4. It covers affiduoufly feventeen days, and all the chicks burft the fhell the fame day.

The young cannot fly till after a month, and yet they retire early with their parents, and often before the greater terns. They are feen flying along the Seine and the Loire at the time of their paffage. Their flight refembles that of the greater terns; they are even continually in the air: they fly oftener fkimming the furface of the water, and rife very high, and with great rapidity.

The BLACK GUIFETTE;
or, the SCARE-CROW.

FOURTH SPECIES.

Sterna Fiffipes. Linn. and Gmel.
Sterna Nigra. Briff.
Larus Niger. Gefn. Aldrov. Johnft. Ray, &c.
The Scare Crow. Will.
The Black Tern. Penn. and Lath.

So much does this bird refemble the preceding, that in Picardy it has been ftiled the *Black Guifette.* The name of *fcare-crow (epou-*

X 3   *ventail)*

ventail) it probably received from the dark cinereous caſt of its head, neck, and body: its
wings only are of a handſome gray, which is the
common garb of the ſea-ſwallows.   It is nearly
as large as the common guifette : its bill is
black, and its ſmall legs are of a dull red.   The
male is diſtinguiſhed by a white ſpot placed
under the throat.

Theſe birds have nothing mournful but their
plumage, for they are very cheerful, fly inceſ-
ſantly, and, like the other ſwallows, make a
thouſand turnings and windings in the air. They
neſtle among the reeds in marſhes, and lay
three or four eggs of a dirty-green, with blackiſh ſpots, that form a zone near the middle *.
They alſo purſue winged inſects, and reſemble
the preceding ſpecies in all their habits †.

* Willughby.
† *Obſervations communicated by M. Baillon, of Montreuil-ſur-mer.*

[A] Specific character of the Black Tern, *Sterna Fiſſipes :* " Its
" body is black, its back cinereous, its belly white, its feet red-
" diſh."   It is very numerous in the Tartarian deſerts.

The GACHET.

FIFTH SPECIES.

Sterna Nigra. Linn. and Gmel.
Sterna Atricapilla. Briff.
The Leffer Sea Swallow. Albin.

A FINE black covers the head, the throat, the neck, and the top of the breaft, like a hood or domino; the back is gray, and the belly white: it is rather larger than the guifettes. The fpecies feems not to be very common on our coafts, but it occurs on thofe of America, where Father Feuillée has defcribed it *, and

* It feems to be indicated by the name *bufc* in the following paffage of the navigator Dampier. " We faw fome *boobies* and *bufcs*, " and at night we took one of the latter: it was different both in " colour and figure from any that I had ever feen; its bill was long " and flender, as in all other birds of this kind; its foot flat like " that of *ducks*; its tail longer, broad, and more forked than that " of *fwallows*; its wings very long; the upper fide of its head " coally black; fmall black ftripes round its eyes, and a pretty " broad white circle which inclofes them on either fide; its craw, " its belly, and the upper fide of its wings, white; but the back and " the under fide of the wings pale black or fmokey. . . . Thefe " birds are found in moft places between the tropics, as well as in " the Eaft Indies, and on the coaft of Brazil; they pafs the night on " land, fo that they never go more than thirty leagues to fea, un- " lefs they are beaten by fome ftorm. When they hover about " veffels, they generally perch at night, and fuffer themfelves to be " taken without ftirring; they make their nefts on the hillocks or " the adjacent fea-rocks."

x 4   observed

obferved that it lays on a bare rock two eggs,
very large for its fize, and mottled with dull
purplifh fpots on a whitifh ground. The fubject examined by this traveller was larger than
the one defcribed by Briffon, who has notwithftanding ranged them together under the name
of *Gachet*.

[A] Specific character of the *Sterna Nigra*: " Its body is hoary,
" its head and bill black, its feet red."

The SEA SWALLOW of the
PHILIPPINES.

SIXTH SPECIES.

Sterna Panayenfis.   Gmel.
The Panayan Tern.   Lath.

SONNERAT found this bird in the ifland of
Panay, one of the Philippines: it is as
large as the common tern, and is perhaps of the
fame fpecies, modified by the influence of climate ; for all the fore fide of its body is white,
the upper fide of the head is fpotted with black ;
and the only difference is, that the wings and
tail are grayifh below, and amber colour above ;
the bill and feet are black.

The SEA SWALLOW OF GREAT ALAR-EXTENT.

SEVENTH SPECIES.

Sterna Fuliginofa. Gmel.
The Egg-Bird. Forfter and Cook.
The Noddy. Hawkefworth and Dampier.
The Sooty Tern. Penn. and Lath.

THOUGH all the fea fwallows have great extent of wings, that character is more remarkable in this fpecies, which is not larger than the common tern, and yet meafures two feet nine inches acrofs the wings. There is a fmall white crefcent on its front; the upper fide of the head and of the tail is a fine black, and all the under fide of the body white; the bill and feet are black. We are indebted to the Vifcount de Querhoënt for the account of this fpecies, which he found at the ifle of Afcenfion. " It " is inconceivable," fays he, " how many fwal- " lows are feen at Afcenfion; the air is fome- " times darkened with them, and the little " plains entirely covered : they are very clamo- " rous, and continually pour forth their harfh, " fhrill cries, exactly like thofe of the white " owl. They are not timorous; they flew over

" my

" my head, and almost touching me : those
" which sat on their nests did not spring as I
" approached, but struck furiously with their
" bill when I attempted to take them. Of more
" than six hundred nests, I saw only three that
" contained two chicks or two eggs : all the
" rest had only one. They were placed on the
" flat ground, near some heaps of stones, and
" all close beside each other. In one part of
" the island, where a flock was settled, I found
" in all the nests the young bird already grown,
" and not a single egg. Next morning I light-
" ed on another colony, where was only one egg
" on which incubation had begun, but no
" chick : this egg, which surprized me by its
" magnitude, is yellowish, with brown spots and
" other spots of pale violet, more crouded on
" the broad end. No doubt these birds have
" several hatches in the year. The young are
" at first covered with a light gray down. When
" caught in the nest, they immediately reject
" the fish from their stomach."

　[A] Specific character of the Sooty Tern, *Sterna Fuliginosa*.
" It is black; its under side, its cheeks, its front, and the shafts of
" all its quills, white."

The GREAT SEA-SWALLOW
of CAYENNE.

EIGHTH SPECIES.

Sterna Cayanensis. Gmel.
The Cayenne Tern. Lath.

THIS species might be stiled the *greatest sea-swallow*, for it exceeds, by two inches, the common sea-swallow of Europe. It is found in Cayenne: and, like most of the preceding, it has all the under side of the body white; a black hood on the back of the head, and the feathers of the mantle fringed on a gray ground with dilute yellowish or rusty.

WE know only these eight species of sea-swallows: and we remove from this family of birds the *cinereous tern* of Brisson, because its *wings are short*; whereas the extent of wings is the chief character by which nature has distinguished them, and is the source of all their other habits.

The TROPIC BIRD.

L'Oiseau du Tropique, *ou* Le Paille-en-Queue *. *Buff.*

WE have seen birds travel from north to south, and with boundless course traverse all the climates of the globe: others we shall view confined to the polar regions, the last children of expiring nature, invaded by the horrors of eternal ice. The present, on the contrary, seems to attend the car of the sun under the burning zone, defined by the tropics †: flying perpetually amidst the tepid zephyrs, without straying beyond the verge of the ecliptic, it informs the navigator of his approach to the flaming barriers of the solar track. Hence it has been called the *Tropic Bird*, because it resides within the limits of the torrid zone.

The most sequestered islands of India and America, situated nearest the equator, seem the

---

* In French *Paille-en-cul*, or *Fetu-en-cul*, (straw-in-arse) and *Queu-de-fleche* (arrow-tail): the Dutch name *Pylstaart*, and the Spanish *Rabo-de-junco*, signify the same.

† Probably in this view Linnæus has given it the poetical name of Phaeton, *Phaëton Æthereus.*

favourite

THE COMMON TROPIC BIRD.

favourite haunts of thefe birds; fuch as that of
Afcenfion, St. Helena, Rodrigue, and ifles of
France and Bourbon. In the vaft expanfe of
the northern Atlantic, they have ftrayed to
Bermudas, which is their fartheft excurfion beyond the bounds of the torrid zone *: they traverfe the whole of this fpace †, and occur again
towards the fouthern limit, where they inhabit
the chain of iflands difcovered by Captain Cook,
the Marquefas, Eafter ifland, the Society and
Friendly iflands. He found them alfo in open
fea near thefe latitudes ‡: for though their appearance may be regarded as the token of the
proximity of land, they ufually rove many hundred leagues, and fometimes venture to prodigious diftances §.

Befides

* " One feldom fees thefe birds except between the tropics and
" at great diftance from land; however, one of the places where
" they multiply is near nine degrees beyond the tropic of Cancer;
" I mean the Bermudas, where thefe birds breed in the clefts of
" the high rocks that gird thefe iflands." *Catefby.*

† The Tropic Birds are found in the greater and the leffer Antilles. See *Dutertre, Labat, Rochefort,* &c.—" In going by fea
" from Fort St. Peter to Fort Royal, in Martinico, diftant feven
" leagues, we obferve lofty cliffs that environ the ifland; in the
" holes of thefe rocks the Tropic Birds hatch." *Remark of M. de
la Borde, king's phyfician at Cayenne.*

‡ The ifland which Tafman difcovered in 22° 36′ latitude fouth,
received the name of *Pylftaart.*

§ " We faw a Tropic Bird in 20° latitude north, and 336° lon-
" gitude. I was furprized to find them at fuch a great diftance
" from land. Our captain, who had made feveral voyages to
" America, obferving my furprize, affured me that thefe birds left
" the iflands in the morning to earn their fubfiftence on the vaft
" ocean,

Befides its powerful and rapid flight, the broad and entirely palmated feet of the Tropic Bird enable it, when fatigued with its diftant journies, to reft on the furface of the water *. Its toes are connected by a membrane as in the cormorants, the boobies, the frigates, which it refembles by this character, and alfo by the habit of perching on trees †. Yet it is more analogous to the fea-fwallows than to any of thefe birds: like them, it has long wings which crofs on the tail when in a ftate of repofe: its bill too is fhaped like theirs, though ftronger, thicker, and flightly indented on the edges.

"ocean, and returned in the evening to their quarters; in fhort, "reckoning foutherly, they muft have been about 500 leagues "from thefe iflands." *Feuillée*, Obferv. 1725, p. 170.

"In 27° 4′ latitude fouth, and 103° 30′ longitude weft, in the "firft days of March, we faw Tropic Birds." *Cook.*—"We faw "man-of-war birds, gulls, and Tropic Birds, which we believed to "come from St. Matthew or Afcenfion, which we had left behind "us." *Id*—"On the 22d of May, 1767, we were by obfervation "in 111° longitude weft, and 20° 18′ latitude fouth; the fame day "we faw bonettoes, dolphins, and Tropic Birds." *Wallis.*—"Be- "ing in 20° 52′ latitude fouth, and 115° 30′ longitude weft, we "caught for the firft time two bonettoes, and we faw feveral; we "faw alfo feveral Tropic Birds." *Byron.*

"In 18 degrees fouth latitude, on the meridian of Juan Fer- "nandez, running eaftward, we faw a number of Tropic Birds." *Le Maire.*—"In 29° latitude fouth, and about 133° longitude weft, "we faw the firft Tropic Bird." *Cook.*

* Labat believes that they even fleep on the water.

† "During three months which I paffed at Port Louis in the "ifle of France, I never obferved any fea-bird except fome Tropic "Birds, which croffed the roads in their way to the woods." *Remarks made by the Vifcount de Querhoënt, on board his Majefty's fhip the Victory, in* 1773 *and* 1774.

It

It is nearly as large as a common pigeon. The fine white of its plumage would alone fuffice to diftinguifh it; but its moft ftriking character is a long double fhaft, which appears like a ftraw fixed into the tail, whence its name in French *. This is formed by the production of the two middle quills of the tail, which is extremely fhort; they are almoft naked, edged only with very narrow webs, and they extend twenty-two or twenty-four inches. Often they are of unequal length, and fometimes only one is feen; which may be owing to fome accident, or to moulting: for in that feafon they drop it, and then the inhabitants of Otaheite and the neighbouring iflands gather thefe long feathers in their woods, whither thefe birds come to repofe at night †; the iflanders weave them into tufts and chaplets for their warriors ‡. The Caribs thruft them through the *feptum* of the nofe, to look handfomer or more ferocious §.

We may readily fuppofe, that a bird whofe flight is fo free, fo lofty, fo vaft, cannot be re-

* *Paille-en-Queue.*

† " As we fet out before fun-rife, Tahea and his brother, who " accompanied us, took fea-fwallows which were fleeping on the " bufhes along the road; they told us, that many water fowls came " to repofe on the mountains after flying the whole day at fea in " queft of food, and that the Tropic Bird in particular repaired to " thefe retreats. The long feathers of its tail, which it fheds an- " nually, are commonly met with on the ground, and the natives " are eager to find them." *Forfter.*

‡ *Idem.*

§ Dutertre.

conciled

conciled to captivity *. Its ſhort legs placed behind render it as heavy and aukward on the ground as it is nimble and active in the air. Sometimes the Tropic Birds, ſpent by the bluſtering of ſtorms, alight on ſhips' maſts, and ſuffer themſelves to be taken with the hand †. Leguat, the navigator, ſpeaks of a diverting conteſt between them and his ſailors, whoſe caps they ſnatched off ‡.

The Tropic Birds have been divided into two or three kinds, which ſeem to be only varieties nearly allied to the common ſtock. We proceed to enumerate theſe, without pretending that they are ſpecifically different.

* " I kept a long time a young Tropic Bird; I was obliged, " though it was conſiderably grown, to open its bill to make it " ſwallow food; it would never eat without aſſiſtance. As much as " theſe birds are nimble on wing, they are heavy and ſtupid in the " cage. As their legs are very ſhort, all their motions are con- " ſtrained: mine ſlept almoſt the whole day." *Remarks made at the iſle of France, by the Viſcount de Querhoënt.*

† Hiſt. Univer. des Voyages, par Montfraiſier; Paris, 1707, *p.* 17.

‡ " Theſe birds annoyed us in a ſingular manner; they ſurprized " us behind, and ſnatched the caps from our heads; and theſe at- " tacks were ſo frequent and ſo troubleſome, that we were obliged " to hold ſticks conſtantly in our hands for defence. We prevented " them ſometimes, when we ſaw before us their ſhadow the moment " they were about to make their aim. We could never underſtand " what uſe our caps could be to them, or what they did with thoſe " which they had carried off." *Voyages & Aventures de Francis Leguat; Amſterdam,* 1708. *tom.* i. *p.* 107.

The GREAT TROPIC BIRD.

FIRST SPECIES.

Phaeton Æthereus. Linn. and Gmel.
Lepturus. Briff.
Avis Tropicorum. Ray and Will.
Plancus Tropicus. Kleln.

THIS exceeds the bulk of a large dove-houfe pigeon; its fhafts are near two feet long; all its plumage is white, with little broken black lines above the back, and a black ftreak, in fafhion of a horfe-fhoe, inclofes the eye at the inner corners; the bill and feet are red. It is found in the ifland of Rodrigue, and in that of Afcenfion, and at Cayenne; and feems the largeft of the genus.

[A] Specific character of the *Phaeton Æthereus :* " It is white; " its back, its rump, and the leffer coverts of its wings, ftreaked with " black; its two middle tail-quills black at the bafe; its bill red." It fometimes roves immenfe diftances beyond the tropics: Linnæus mentions the latitude of $47\frac{1}{2}$ degrees as the limit; and I myfelf faw one nearly in that parallel, between the Bank of Newfoundland and the Channel. Linnæus adds, that the Tropic Bird feeds on macka-rels, dolphins, and fharks (I fuppofe he means the dead carcafes that fometimes float on the furface).

The LITTLE TROPIC BIRD.

SECOND SPECIES.

Phaeton Æthereus. var. 1. Linn. and Gmel.
Lepturus Candidus. Briff.
Alcyon Media Alba. Brown.

THIS is fcarce equal in fize to a common fmall pigeon. Like the preceding, it has the horfe-fhoe about the eye, and is befides fpotted with black on the feathers of the wings neareft the body, and on the great quills: all the reft of its plumage is white, and alfo its long fhafts. The edges of the bill, which in the great Tropic Bird were ferrated with reflected incifures, are much lefs fo in this. It vents at intervals a fmall cry, *chiric, chiric,* and makes its neft in the holes of craggy rocks: it lays two eggs, according to Father Feuillée, which are bluifh, and rather larger than thofe of a pigeon.

On comparing feveral individuals of this fecond kind, in fome we remarked reddifh or fulvous tints on the white ground of the plumage. This variation proceeds, we prefume, from the tender age; and to the fame caufe we would attribute the *fulvous* caft, defcribed by

8   Briffon,

Briffon *, efpecially as he reprefents that bird as
rather fmaller than his *white* one. We alfo
perceived confiderable diverfity in the bulk of
thefe birds. Many travellers have affured us,
that the young ones are not pure white, but
potted or ftained with brown or blackifh : they
differ alfo, becaufe their fhafts and feet, inftead
of being red, are pale blue. We muft, however,
obferve, that though Catefby affirms, in general,
that thefe birds have their bill and legs red, this
is not invariably true, but of the preceding fpecies and of the following ; for in this fpecies,
which is the moft common in the ifle of France,
the bill is yellowifh, like horn, and the legs are
black.

* " Tawny white ; a bar above the eyes ; the fcapular feathers
" near their extremity, and a ftripe above the wings, black ; the tail-
" quills tawny white, their fhafts blackifh at the origin." *Lepturus
Fulvus.*

The RED-SHAFTED
TROPIC BIRD.

THIRD SPECIES.

Phaeton Phœnicurus. Linn. and Gmel.
The Red-tailed Tropic Bird. Lath.

THE two long fhafts of the tail are of the
fame red with the bill ; the reft of the
plumage is white, except fome black fpots on

Y 2   the

the wing near the back, and a black horse-shoe which environs the eye. The Viscount de Querhoënt was so obliging as to communicate the following note on this bird, which he observed at the isle of France. " The Red-shafted " Tropic Bird breeds in this island, as well as " the common tropic bird; the latter in the " hollow trees of the principal island, the for- " mer in the cavities of the small neighbouring " islets. The Red-shafted Tropic Bird is scarce " ever seen on land; and, except in the season " of courtship, the common tropic bird seldom " comes ashore. They live by fishing at large, " and come to repose on the small isle of *Coin-* " *de-Mire*, which is two leagues from the isle of " France, and is the haunt of many other sea " birds. It was in September and October that " I found the nests of the tropic birds: each " contained only two eggs of a yellowish white, " marked with rusty spots. I was assured, that " no more than one egg is found in the nest of " the great tropic bird: and none of the species " seem to be numerous *."

None

---

* " While I was seeking for them, chance led me to be spectator " of a fight between the *martins* and the *Tropic Birds :* having been " directed into a wood, where I was told that these birds had set- " tled, I sat myself down at some distance from the tree marked, " where I saw several martins collect: a short while after the Tro- " pic Bird arrived to enter its hole; and the martins rushed upon it " and attacked it on all sides, and though it has a very strong bill, " it was obliged to flee; it made several attempts, which were not
" more

None of these three species or varieties, which we have just described, appears attached to any particular spot; often the two first or the two last are found together; and the Viscount de Querhoënt says, that he saw all the three collected at the island of Ascension.

" more fortunate, though assisted at length by its mate. The martins,
" proud of their victory, did not quit the tree, and were on it when
" I left them." *Sequel of the Viscount de Querhoënt's note.*

[A] Specific character of the *Phaeton Phœnicurus:* " It is of a
" very pale rose-colour; its bill, and its two middle tail-quills, are
" red."

The BOOBIES.

LES FOUS *. *Buff.*

IN every well organized being, inſtinct diſplays itſelf by a chain of conſiſtent habits, which all tend to its preſervation; and this internal ſenſe directs them to ſhun what is hurtful, and to ſeek what may contribute to the ſupport, and even the enjoyment, of life. The birds that we are now to ſurvey, have received from nature only half that faculty: large and ſtrong, armed with a firm bill, provided with long wings, and with feet completely and broadly palmated, they are fitted to exerciſe their powers both in the air and in the water, they are invited to act and to live; yet they ſeem ignorant what exertions they ſhould make, or what precautions they ſhould obſerve, to eſcape that death which perpetually threatens them. Though diffuſed from one end of the world to the other, from the ſeas of the north to thoſe of the ſouth, they have no where learnt to diſtinguiſh their moſt dangerous enemy: the ſight of man does not intimidate or diſcompoſe them. They ſuffer themſelves to

* By the Portugueſe ſettlers in India, they are called *Paxaros Bobos,* or the fooliſh birds.

be

be taken, not only at fea on the fhips' yards *,
but alfo at land, on the iflets and coafts, where
they may be felled by blows with a ftick, in
great numbers, one after another, and yet the
ftupid flock will make no effort to efcape †.
This infenfibility to danger proceeds neither
from refolution nor courage: fince they can neither refift nor defend, ftill lefs can they attack,
though their ftrength and their armour might
render them formidable ‡.   It originates, therefore, from ftupidity and imbecility.

As

---

* Thefe birds are called *Boobies* (fous) becaufe of their great
ftupidity, their filly-afpect, and their habit of continually fhaking
the head and fhivering when alighted on fhips' yards, or other parts,
where they fuffer themfelves to be taken by the hand. *Feuillée.*—
If the Booby fees a fhip, either in open fea or near land, it will come
to perch on the mafts; and fometimes, if a perfon ftretches out his
hand, the bird will alight upon it. In my voyage to the iflands,
there was one which paffed fo often over my head, that I transfixed
it with a half-pike. *Dutertre.*—Thefe birds are not at all fhy, either
on land or at fea; they approach a veffel without feeming to fear
any thing, when they chance to come in the way: the report of a
fowling-piece, or any other noife, will not deter them. I have fometimes feen one of thefe folitary Boobies come to rove about the fhip
at evening, and to alight on the yards, where the failors caught
them without their fhewing the fmalleft inclination to efcape. *Obfervations communicated by M. de la Borde, King's phyfician at Cayenne.*
See alfo Labat, *Nouveau Voyage aux îles de l'Amerique: Paris,* 1722,
tom. vi. *p.* 481.   Leguat, *tom.* i. *p.* 196.

† It is a very filly bird, and will hardly get out of peoples' way.
Dampier.—In this ifland of Afcenfion, the Boobies are fo numerous,
that our failors killed five or fix with one blow of a ftick. *Gennes.*
—Our foldiers killed an aftonifhing quantity of them at the fame
ifland of Afcenfion. *Vifcount de Querhoënt.*

‡ The Boobies are certain birds fo called, becaufe they fuffer
themfelves to be caught by the hand: they pafs the day on the rocks,

As the mental powers and the moral qualities of animals are derived from their conftitution, we muft attribute the exceffive fluggifhnefs
and helplefs fecurity of the Boobies to fome phy-
fical caufe; and this, moft probably, is the difficulty of putting their long wings in motion *.

But man is not their only foe; their want of
courage expofes them to another enemy, which
perpetually harraffes them. This is the frigat,
or man-of-war bird. It rufhes upon the Boobies which it defcries, purfues them without intermiffion, and obliges them, by blows with its
wings and its bill, to furrender their prey, which
it inftantly feizes and fwallows † : for the filly,

which they never leave but when they go a-fifhing; in the evening
they retire to the trees, and, after they are once perched, I am perfuaded they would not quit, though thefe were fet on fire: and they
will all fuffer themfelves to be taken without ftirring from the fpot;
however they try to do their beft in defence with their bill, but they
cannot hurt a perfon. *Hiftory of the Buccaneers*, 1686.

* We fhall fee that the frigat itfelf, notwithftanding its vigorous
wing, has the fame difficulty in taking its flight.

† I had the pleafure to fee the frigats give chafe to the Boobies:
when they retire in bodies at evening from the labours of the day,
the frigats watch their return, and, rufhing on, oblige them to
fcream for affiftance, and to difgorge fome fifh, which they carry to
their young. *Feuille.*—The Boobies repair at night to repofe on
the ifland of Rodrigue, and the frigats, which are large birds, and fo
called becaufe of the rapidity of their flight, wait for them every
evening on the tops of the trees; they rife very high, and dart
down upon them like a hawk upon his prey, not to kill them, but
to make them difgorge; the Booby, ftruck in this way by the frigat, throws up a fifh, which the latter fnatches in the air: often
the Booby fcreams, and difcovers a reluctance to part with its
booty; but the frigat fcorns its cries, and, rifing again, comes down
with fuch a blow as to ftun the poor bird, and compel an immediate
furrender. *Leguat.*

cowardly

cowardly Boobies difgorge at the firft attack, and
return to feek new prey, which they often lofe
by a fecond piracy *.

The Boobies hover above the furface of the
water, fearce moving their wings, and drop on
the fifh the inftant it emerges †. Their flight,
though rapid and well fupported, is greatly inferior to that of the frigat. Accordingly, they
do not roam fo far, and their appearance is regarded by navigators as a pretty certain fign of
the nearnefs of fome land ‡. Yet feveral of
thefe

* Catefby defcribes fomewhat differently the fkirmifhes of the
Booby and its enemy, which he calls the *pirate*. " The latter "
fays he, " fubfifts entirely on the fpoils of others, and particularly
" of the Booby. As foon as the pirate perceives that it has caught
" a fifh, he flies furioufly againft it, and obliges it to dive under wa-
" ter for fafety ; the pirate not being able to follow it, hovers above
" the water till the Booby is obliged to emerge for refpiration, and
" then attacks it again while fpent and breathlefs, and compels it
" to furrender its fifh ; it now returns to its labours, and has to fuf-
" fer frefh attacks from its indefatigable enemy."

† Ray.

‡ The Boobies do not go very far to fea, and feldom lofe fight of
land. *Forfter.*—A few days after our departure from Java, we faw
Boobies about the fhip for feveral nights together; and as thefe birds
go to rooft on land in the evening, we conjectured that there was fome
ifland near us ; perhaps it was the ifland of Selam, whofe name and
pofition are very differently marked on the charts. *Cook.*—Our
latitude was 24° 28′ (on the 21ft May 1770, near New Holland);
we had found on the preceding days feveral fea birds, called Boobies,
but we had not that fight to-day. On the night of the 21ft, there
paffed near the fhip a fmall flock flying to the north-weft;
and in the morning, from an hour before fun-rife to half an hour
after, there were continual flights that came from the north north-
weft, and difappeared toward the fouth fouth-eaft; we faw none that
took

thefe birds frequent our northern coafts *, and
occur in the remoteft and moft fequeftered
iflands in the midft of the ocean †. There they
live in companies, with the gulls, the tropic birds,
&c. and the frigat, their inveterate foe, has followed them to their retreats.

Dampier gives a curious account of the hoftilities between the man-of-war birds and the
Boobies, in the Alcrane iflands, on the coaft of
Yucatan. " Thefe birds were crowded fo thick
" that I could not," he fays, " pafs their haunt
" without being incommoded by their pecking.
" I obferved that they were ranged in pairs,
" which made me prefume that they were male
" and female. When I ftruck them, fome flew

took another direction, which led us to fuppofe, that at the bottom
of a deep bay lying fouth of us, there was a lagoon, or fhallow river, whither thefe birds repaired to feek their food during the day,
and that on the north of us there was fituated fome ifland to which
they retired. *Cook.*—*Note,* We muft confefs that fome voyagers,
and among others Father Feuillée, fay that Boobies are found feveral hundred leagues at fea; and that Captain Cook himfelf feems to
reckon them, at leaft in certain circumftances, as more certain tokens of the proximity of land than the frigats, with which he claffes
them in the following paffage. " The weather was pleafant, and
" every day we faw fome of the birds which are efteemed to be
" figns of nearnefs of land, fuch as Boobies, frigats, tropic birds,
" and gulls. We believed that they came from the ifland of St.
" Matthew or Afcenfion, which we had left pretty near us."
* See the article of the Gannet.
† At Rodrigue, *Leguat:* at Afcenfion, *Cook:* at the Calamiane
iflands, and at Timor, *Gemelli Careri:* at Sabuda in New Guinea,
and at New Holland, *Dampier:* in all the iflands fcattered under
the fouthern tropic, *Forfter:* in the Great Antilles, *Feuillée, Labat,
Dutertre,* &c.: in the Bay of Campeachy, *Dampier.*

" away,

" away, but the greater number remained, and
" would not ftir for all I could do to rouze them.
" I remarked alfo, that the man-of-war birds
" and the Boobies always placed fentinels over
" their young, efpecially when they went to fea
" for provifion. Of the man-of-war birds, many
" were fick or maimed, and feemed unfit to
" procure their fubfiftence. They lived not
" with the reft of their kind, whether they
" were expelled from the focicty, or had fepa-
" rated from choice : thefe were difperfed in
" different places, probably that they might have
" a better opportunity of pillaging. I once faw
" more than twenty on one of the iflands
" fally out from time to time into the open
" country, to carry off booty, and they returned
" again almoft immediately. When one fur-
" prized a young Booby that had no guard, he
" gave it a violent peck on the back to make it
" difgorge, which it did inftantly : it caft up
" one or two fifh about the bulk of one's hand,
" which the old man-of-war bird fwallowed ftill
" more haftily. The vigorous ones play the
" fame game with the old Boobies which they
" find at fea. I faw one myfelf which flew right
" againft a Booby, and with one ftroke of its
" bill, made him deliver up a fifh which he had
" juft fwallowed. The man-of-war bird darted
" fo rapidly as to catch it in the air before it
" could fall into the water."

The Boobies refemble moft the cormorants in
their

their fhape and organization, except that their bill is not terminated in a hook, but in a point flightly curved : they differ alfo, becaufe their tail projects not beyond their wings. They have their toes connected by a fingle piece of membrane ; the nail of the mid-one is ferrated on the infide : their eyes are encircled by a naked fkin ; their bill is ftraight, conical, and fomewhat hooked at the end, and the fides are finely indented ; the noftrils are not apparent, and their place is occupied only by two hollow channels. But the moft remarkable property of the bill is, that the upper mandible is articulated, as it were, and formed of three pieces joined by two futures ; the firft is traced near the point, which therefore appears like a detached nail ; the fecond is fituated at the root of the bill near the head, which enables the bird to raife the tip of its upper mandible two inches, without opening the bill *.

Thefe birds utter a loud cry, partaking of that of the raven and of the goofe ; and this is heard particularly when they are purfued by the frigat, or when, affembled together, they are feized by fome fudden panic †. In flying they ftretch out the neck, and difplay the tail. They cannot begin

---

* " What is moft remarkable in thefe birds, the upper mandible, " two inches below the mouth, is jointed in fuch manner that it can " rife two inches above the lower mandible, without the bill being " opened." *Catefby*.

† " We had been hunting goats at night (in the ifland of Af- " cenfion) ; the reports of the piece which we fired had fright- " ened

THE COMMON BOOBY.

begin their motion but from fome lofty ftation, and therefore they perch like cormorants. Dampier remarks, that in the ifle of Aves they breed on trees, though in other places they neftle on the ground, and always a number in the fame haunt; for a community, not of inftinct but of weaknefs, feems to collect them together. They lay only one or two eggs. The young ones continue long covered, for the moft part, by a very foft and white down.—The other particulars will beft appear in the enumeration of their fpecies.

" ened the Boobies in the neighbourhood: they all fcreamed toge-
" ther, and the reft replied at fhort diftances, which made a hideous
" din." *Note communicated by the Vifcount de Querhoënt.*

The COMMON BOOBY.

FIRST SPECIES

Pelecanus-Sula. Linn. and Gmel.
Sula. Briff.
Plancus Morus. Klein.

THIS bird, which feems to be moft common in the Antilles, is of a middle fize between the duck and the goofe: its length, from the end of the bill to that of the tail, is two feet five inches, and a foot eleven inches to the extremities of the nails: its bill is four inches and

and a half, and its tail is near ten; the naked
fkin which encircles the eye is yellow, and fo is
the bafe of the bill, whofe point is brown; the
legs are ftraw-coloured *; the belly is white,
and all the reft of the plumage is brown cinereous.

Simple as this garb is, it is infufficient, as
Catefby obferves, to characterize the fpecies,
fo many are the individual varieties which it
contains. " I obferved," fays he, " one that had
" a white belly and a brown back; another,
" whofe breaft and belly were white, and others
" which were entirely brown." Some travellers
feem to denominate this fpecies the *fulvous
bird* †. The flefh is black, and has a marfhy
flavour; yet the failors and adventurers of the
Antilles often fed on it. Dampier relates, that
a fmall French fleet, being caft on the ifle of
Aves, partly fubfifted on thefe birds, and made
fuch confumption of them, that the number
there has fince been much diminifhed.

* Catefby.
† The birds which the French in the Antilles call *fauves*, becaufe
of the colour of their back, are white under the belly; they are of the
bulk of a water-hen, but are ufually fo lean that their plumage is the
only part of them the leaft valuable: they have the feet of ducks, and
the pointed bill of woodcocks; they live on fmall fifh, like the frigats, but they are the moft ftupid of the birds, either at fea or on
land, in the Antilles; fince, whether that they eafily tire on wing,
or that they take the fhips for floating rocks, as foon as they perceive one, efpecially if towards night, they immediately come to
alight upon it, and are fo filly as to fuffer themfelves to be taken by
the hand. *Hiftoire Naturelle & Morale des Antilles; Rotterdam,* 1658,
p. 148.

They

They are found in great numbers not only on
the iſle of Aves, but in that of Remire, and
eſpecially at the *Grand-Connétable*, a rock ſhaped
like a ſugar-loaf, riſing apart in the ſea, within
ſight of Cayenne *. Multitudes alſo occur on
the iſlets which lie along the ſhores of New
Spain and Caracca †. And the ſame ſpecies
ſeems to be met with on the coaſt of Brazil ‡,
and on the Bahama iſlands, where, it is aſſerted, they lay every month of the year two or
three eggs, or ſometimes only one, on the naked
rocks §.

* Barrere, *France Equinoxiale, p.* 122.

† What makes theſe birds and many others ſo extremely numerous on theſe ſhores, is the incredible ſwarms of fiſh which attract
them: a perſon can ſcarce let down into the water a line with
twenty or thirty hooks, but he finds, on drawing it up, a fiſh hanging from each.

‡ On theſe iſlands (of St. Anne, on the coaſt of Brazil) numbers
are found of large birds, called Boobies *(fous)* becauſe they allow
themſelves to be eaſily caught: in a ſhort time we took two dozen... Their plumage is gray; they are ſkinned like hares. *Lettres
Edifiantes,* xv. *Recueil, p.* 339.

§ Cateſby.

[A] Specific character of the Booby, *Pelecanus-Sula:* "Its tail
" is wedge-ſhaped, its body whitiſh, its primary wing-quills black
" at the tip, its face red."

The WHITE BOOBY.

SECOND SPECIES.

Pelecanus-Piscator. Linn. and Gmel.
Sula Candida. Briff.
The Leffer Gannet. Lath.

WE have remarked, that there is much diverfity of white and brown in the preceding fpecies, yet we cannot clafs this with it;
the more fo as Dutertre, who faw both alive,
diftinguifhed them from one another. They
are indeed very different, fince what is white in
the one is brown in the other; viz. the back,
the neck, and the head, which is befides rather
fmaller. It appears alfo to be lefs ftupid; it feldom perches on trees, and ftill lefs does it
fuffer itfelf to be caught on the fhips' yards;
yet it inhabits the fame places with the preceding, and both are found on the ifland of
Afcenfion. " There are," fays the Vifcount de
Querhoënt, " in this ifland, thoufands of common
" Boobies; the white are lefs numerous; both
" kinds are feen perched upon heaps of ftones, ge-
" nerally in pairs. They are found at all hours, and
" will never ftir till hunger obliges them to fifh.
" Their general refort is on the windward fide of the
" ifland.

" ifland. They may be approached in broad
" day, and caught even by the hand. There are
" Boobies alfo which differ from the preceding :
" when at fea, in the latitude of 10° 36' north,
" we faw fome whofe head was entirely
" black *."

* Captain Cook found White Boobies on Norfolk ifland.

[A] Specific character of the Leffer Gannet, *Pelecanus Pifcator :*
" Its tail is wedge-fhaped ; its body white ; all its wing-quills
" black ; its face red."

The GREAT BOOBY.

THIRD SPECIES.

Pelecanus Baffanus, var. 1. Linn. and Gmel.
Sula Major. Briff.

THIS bird is the largeft of its genus, being
equal to the goofe, and its wings meafur-
ing fix feet acrofs : its plumage is deep brown,
fprinkled with fmall white fpots on the head,
with broader ones on the breaft, and with others
ftill broader on the back ; the belly is dirty
white. The colours are more vivid in the male
than in the female.

This large bird is found on the coafts of
Florida, and on the great rivers of that country.

" It dives," fays Catefby, " and remains a con-
" fiderable time under water, where I imagine it
" chances on fharks and other voracious fifh,
" which often maim or deftroy it; for I feveral
" times found thefe birds wounded or dead on
" the beach."

An individual of this fpecies was taken in the
neighbourhood of the city of Eu, on the 18th
of October, 1772.   No doubt it had been furprized far at fea by rough weather, and driven
by the violence of the wind upon our coafts.
The perfon who found it had only to throw his
coat over it : it was kept fome time ; at firft it
would not ftoop to take a fifh, but required it
to be held as high as its bill.   It fat always
fquat, and was averfe to motion ; but after being accuftomed to live on land, it walked and
became familiar; it even importunately followed
its mafter, making at intervals a fhrill raucous
cry *.

* Extract of a letter from the Abbe Vincent, profeffor in the
college of the city of Eu, inferted in the *Journal de Phyfique* for
June, 1773.

The LITTLE BOOBY.

FOURTH SPECIES.

Pelecanus Parvus. Gmel.
The Leſſer Booby. Lath.

THIS is the leaſt of the Boobies known: its length, from the end of the bill to that of the tail, is ſcarcely a foot and half; the throat, the ſtomach, and the belly, are white, and all the reſt of the plumage is blackiſh. It was ſent to us from Cayenne.

The LITTLE BROWN BOOBY.

FIFTH SPECIES.

Pelecanus Fiber. Linn. and Gmel.
Fiber Marinus. Feuillée.
Larus Piſcator Cinereus. Klein.
Sula Fuſca. Briſſ.

THIS bird differs from the preceding, being entirely brown; and though it is alſo larger, it equals not the common Booby. We

z 2   therefore

therefore range thefe fpecies feparately, till new obfervations inform us whether they ought to be joined. Both of them inhabit the fame places, and particularly Cayenne and the Caribbee iflands.

[A] Specific character of the *Pelecanus Fiber :* " Its tail is " wedge-fhaped; its body dufkifh; all its wing-quills blackifh; its " face red."

The SPOTTED BOOBY.

SIXTH SPECIES.

Pelecanus Maculatus. Gmel.

THE colours and bulk of this bird might re-fer it to the third fpecies, did it not differ in the exceffive fhortnefs of its wings. Indeed, we fhould almoft doubt, whether it belonged to the Boobies, but for the characters of its bill and feet. It is equal to the great diver, and as in it, the ground of the plumage is blackifh brown, wholly fpotted with white, more delicately on the head, and broader on the back and wings; the ftomach and belly are waved with brownifh, on a white ground.

[A] Specific character of the *Pelecanus Maculatus :* " It is " brown fpotted with white; below white, waved and fpotted with " brown; its bill, its wing-quills, its tail, and its feet brown."

The G A N N E T.

Le Fou de Bassan. *Buff.*

Pelecanus Baffanus. Linn. and Gmel.
Sula Baffana. Briff.
Anfer Baffanus. Sibbald, Ray, Charleton, &c.
Anfer Raffanus, vel Scottcus. Gefner, Aldrov. &c.
Sula Hoieri. Clufius and Will.
The Solan Goofe *. Will. and Alb.

THE Bafs-ifle is a ftupendous rock in the
Firth of Forth, not far from Edinburgh.
It is the refort of thefe large and beautiful birds,
which have been reckoned peculiar to it † : but
Clufius and Sibbald affure us, that it occurs alfo
on the Craig of Ailfa ‡ in the Firth of Clyde,
and in the Hebrides § and the Feroe iflands ||.

This bird is as large as a goofe ; it is near
three feet long, and more than five feet acrofs

---

* In Norwegian *Sule, Hav-Sule.*

† Ray.

‡ Sibbald.

§ Some perfons affure us, that thefe Boobies are at times driven
by adverfe winds on the coafts of Brittany, and that one was ıeen
even in the vicinity of Paris.

|| Hector Boece, in his defcription of Scotland, fays, that thefe
birds alfo neftle on the Hebrides ; but what he adds, that for this
purpofe they bring as much wood as to fupply the inhabitants,
feems fabulous ; efpecially as the Gannets of the Bafs lay, like the
other boobies of America, on the naked rock.

the

the wings : it is entirely white, except the pri-
maries of the wing, which are brown or blackifh, and the back of the head, which is tinged
with yellow * : the cere is of a fine blue, and
alfo the bill, which extends fix inches, and opens
fo wide as to admit a large mackerel ; nor does
this enormous morfel always fatisfy its voracity.
M. Baillon fent us a Gannet that was taken in
open fea, and which had choaked itfelf in fwallowing a very large fifh †. Near the Bafs, and
at the Hebrides, they fubfift generally on herrings. Their flefh contracts a fifhy tafte : but
the young ones are always very fat ‡ ; and
perfons defcend among the crags to rob the
nefts §. The old ones might eafily be felled with
fticks or ftones ||, but they are unfit for eat-

* I am inclined to believe that this is a mark of age ; this yellow fpot is of the fame nature with that on the lower part of the
neck of the fpoon-bills : I have feen fome wherein it was golden ;
the fame thing happens to white hens, which turn yellow as they
grow old. *Note communicated by M. Baillon.*—Ray is of the fame
opinion ; and Willughby relates, that the young ones are at firft
marked with brown or blackifh on the back.

† Sent from Montreuil-fur-mer, by M. Baillon, December,
1777. The ftory related by Gefner is fabulous, that, on feeing another fifh, it difgorges the one which it had juft fwallowed.

‡ Gefner fays, that the Scotch make an excellent kind of oint-
ment of the fat of thefe birds.

§ "The art of cookery," fays Sir Robert Sibbald, " cannot form
" a difh of fuch delicate flavour, and combining the taftes of fifh
" and flefh, as a roafted Solan goofe ; and the young grown ones
" are defervedly efteemed delicacies with us, and fell at a high
" price."

|| *Note communicated by James Bruce, Efq; 30th of May, 1774.*

ing.

ing *. They are as filly as the other boobies †.

They breed in all the clefts of the Bafs, and
lay but one egg ‡. The people fay that they
hatch it ftanding on one foot §, a notion fuggefted
probably by the breadth of its fole ‖. It is
widely palmated, and the middle and outer toes
are each near four inches long, and all the four
are connected by an entire piece of membrane :
the fkin does not adhere to the body ; it is connected to it only by fmall bundles of fibres
placed at equal diftances, fuch as one or two
inches, and capable of being extended as much ;
fo that the fkin may be drawn out like a membrane, and inflated like a bladder. The bird,
no doubt, thus fwells itfelf to diminifh its fpecific gravity, and facilitate its flight ; yet no
ducts can be traced from the thorax to the cu-
ticle : but perhaps the air penetrates it through
the cellular texture, as in many other birds.
This obfervation, which will certainly apply to

---

* " It is a bird exceffively fœtid ; in preparing the fpecimen
" for my cabinet, my hands retained the fmell more than a fort-
" night ; and though I dipt the fkin in alkaline lye, and feveral
" times fumigated it with fulphur in the courfe of two years, its
" odour ftill adheres to it." *Note communicated by M. Baillon.*

† *In domibus nutrita ftupidiffima avis.* Sibbald.

‡ Sibbald.

§ Mr. Bruce.

‖ Hence, it is alledged, they received the name of *Sole-an-geefe* ;
but Martin informs us, that this word is of Irifh or Erfe derivation,
and fignifies quick-fighted ; thefe birds being noted for the bright
luftre of their eyes.—*T.*

all

all the species of boobies, was made by M. Daubenton the younger, on a Gannet, sent fresh from the coast of Picardy.

The Gannets arrive in spring on the islands of the north, and retire in autumn *, and advance farther south. Perhaps, if their migrations were well known, it would be found, that they join the other species of boobies on the coasts of Florida; the general rendezvous of all the birds which descend from the boreal regions, and have vigour of wing sufficient to traverse the Atlantic ocean.

* Sibbald.

[A] Specific character of the Gannet, *Pelecanus Bassanus:* " Its " tail is wedge-shaped; its body white; its bill, its primary wing- " quills black; its face blue." It has a small dilatable pouch under its chin, able to contain five or six herrings, which, in the breeding season, it carries to its family. Its legs and toes are black, with a stripe of fine velvet green on the fore part: the tail contains twelve sharp taper quills. The egg is white, and rather smaller than that of a common goose: if it be removed, the bird will lay another; and if this be equally unfortunate, she will even lay a third. The nest is large, and composed of substances that float on the water, as sea-weeds, fog, shavings, &c. It is very probable, that the Gannets attend the progress of the herrings; the fishermen reckon them a sure sign of the approach of the shoal. In December, these birds are frequently seen near Lisbon, diving for sardines, a kind of pilchards. They descend from a vast height, and plunge many fathoms under water. In Scotland they are usually called *Solan-geese*; in Cornwall and Ireland *Gannets*; and in Wales *Gan*. The inhabitants of St. Kilda, we are assured by Martin, take often 22,600 of the young birds annually, besides a prodigious number of eggs. These spoils are the chief subsistance of these hardy islanders, and they store up their provisions in pyramidal stone buildings, covering them over with peat-ashes. The Craig of Ailsa resembles much in appearance the Bass-isle: of the

latter,

2

latter, we have an elegant defcription, by the immortal difcoverer of the circulation of the blood, Dr. Harvey.

I fhall take the liberty of fubjoining a tranflation of it : " There " is a fmall ifland which the Scotch call the *Bafs*, not above a mile " in circuit. In the months of June and July, the furface of this " ifland is fo ftrewed with nefts, and eggs, and young birds, that a " perfon can hardly fet his foot without treading on them. And fo " vaft is the multitude of thofe which fly over head, that like clouds, " they darken the fun and the fky; and fuch is their clangorous " noife, that you can fcarce hear the voice of your companions. " If from the fummit of the lofty precipice, you look towards the " fea which fpreads below, you will perceive, wherever you turn " your eyes, birds innumerable of various kinds, fwimming and " hunting for their prey. If failing round, you furvey the im- " pending cliff, you will fee in every crag and fiffure of the indent- " ed rock, birds of all forts and fizes, which would out-number the " ftars that appear in a clear night. If from a diftance you be- " hold the flocks roving about the ifland, you would imagine them " to be a vaft fwarm of bees."— *De Generat. Animal.* Exer. 2.

The F R I G A T.

LA FREGATE. *Buff.*

Pelecanus Aquilus. Linn. and Gmel.
Fregata. Briff.
Fregata Avis. Ray, Will, &c.
The Man-of-War Bird. Brown, Damp. and Sloane.
The Frigat Bird. Alb. and Penn.
The Frigat Pelican. Lath.

THE fteadinefs and rapidity with which this bird moves through the air, have procured it the name of *Frigat*. It furpaffes all the winged failors in the boldnefs, the vigour, and the extent of its flight; poifed on wings of prodigious length, which fupport it without perceptible motion, it fwims gently through the tranquil air, waiting to dart on its prey with the rapidity of a flafh: but if the atmofphere is embroiled with tempefts, the Frigat, nimble as the wind, afcends above the clouds, and ftretches beyond the region of ftorms *. It journies in all directions, and either mounts upwards or glides horizontally; and it often roams to a diftance of feveral hundred leagues †:
and

* Ray.

† *Idem.*—" There is no bird in the world that flies higher,
" longer, or more eafily, and which roves farther from land. .... It is
" found

THE FRIGAT PELICAN.

and thefe immenfe excurfions are performed by
a fingle flight; and as the day is infufficient, it
purfues its route during the darknefs of the
night, and never halts on the fea, but when invited by the abundance of prey *.

The flying-fifhes, whofe columns are purfued
by the bonettoes, dolphins, &c. when driven to
extremity, fpring out of the water, but efcape
not the Frigats: it is in queft of thefe fifhes
that they roam fo far from the land; they dif-
cern at a vaft diftance † the progrefs of their
phalanxes, which fometimes are fo compacted
as to make a rippling, and to whiten the face of
the ocean. Then the Frigats fhoot with down-
ward flight, and bending along the furface of

" found in the midft of the ocean, three or four hundred leagues
" from land; which fhows its prodigious ftrength and its furprizing
" lightnefs: for it cannot reft on the water like the water-fowl,
" fince its feet are not calculated for fwimming, and its wings are
" fo large, that they require room to begin their motion; if there-
" fore it fell on the water, its efforts would be fruitlefs, and it could
" never rife again. We may hence conclude, that as it is found
" three or four hundred leagues from land, it muft defcribe a track
" of feven or eight hundred leagues before it can halt." Labat,
Nouveaux Voyages aux îles de l'Amerique; *Paris*, 1722, *tom.* vi.

* " In the evening we faw feveral birds called Frigats; at mid-
" night I heard others about the veffel; and at five o'clock in the
" morning we perceived the ifland of Afcenfion." *Wallis.*

† " The dolphins and bonettoes purfued the fhoals of flying fifh,
" as we have obferved in the Atlantic Ocean; while feveral large
" black birds with long wings and a forked tail, ufually called Fri-
" gats, rofe very high in the air, and dafhing down with furpriz-
" ing fwiftnefs on the fifh which they perceived fwimming, never
" failed to ftrike their prey." *Cook.*

the

the water *, they fnatch the fifh, feizing it with
the bill or talons, and often with both at once;
according as it fcuds on the furface, or fprings
into the air.

It is between the tropics only, or a little beyond them †, that we find the Frigat in the feas
of both continents ‡. He maintains a fort of
empire over the birds of the torrid zone: he
obliges many, fuch as the boobies, to provide
for him; and ftriking them with his wing, or
biting them with his hooked bill, he conftrains
them to difgorge their prey, which he inftantly
catches §. The hoftilities which he commits
have

* " Though the Frigat rifes to a vaft height in the air, and
" often beyond the reach of our fight, it notwithftanding defcries
" clearly where the dolphins are in purfuit of the flying fifh: it
" then fhoots down like lightning, not quite to the water, but when
" it has come within ten or twelve fathoms, it makes a great bend,
" and finks gradually till it raze the fea, and catches the little fifh
" either while flying or while in the water, with its bill or its talons,
" and often with both together." *Dutertre.*

† " In 30° 30′ fouth latitude, we began to fee Frigats." *Cook.*—
" In 27° 4′ fouth latitude, and 103° 56 weft longitude, about the
" beginning of March, we met with great numbers of birds, fuch as
" Frigats, tropic birds, &c." *Idem.*

‡ At Ceylon; in the run between Madagafcar and the Maldives;
at the ifland of Afcenfion; at Eafter ifland; at the Marquefas; at
Otaheite, and in all the low iflands of the fouthern Archipelago;
on the coaft of Brazil, where it is called *caripira*; at Caracca; at
the ifle of Aves, and in all the Antilles.

§ " Thefe birds, called Frigats, hunt the boobies; they make
" them rife above the rocks where they are perched, and purfue
" them, ftriking with the ends of their wings; the boobies, the
" better to efcape their enemies, difgorge what fifh they have
" taken; and the Frigats, which want nothing elfe, catch the fpoils

" as

have led failors to beftow on him the appellation of *Man-of-War bird* *. He has the audacity even to fet man at defiance : " On land-
" ing at the ifland of Afcenfion," fays the Vifcount de Querhoënt, " we were furrounded by
" a cloud of Frigats. With a blow of my cane
" I knocked down one, which attempted to
" fnatch a fifh out of my hand: at the fame
" time many of them flew a few feet above the
" kettle which was boiling afhore, and endea-
" voured to carry off the flefh, though a part
" of the fhip's company attended it."

This temerity of the Frigat proceeds as much
from the force of its arms, and the boldnefs of
its flight, as from its voracity. It is fitted by
nature for war : its talons are fharp, its bill terminates in a very pointed hook, its legs are fhort
and ftrong, its flight is rapid, its fight acute :
all thefe qualities feem to mark an analogy to
the eagle, and to conftitute it the tyrant of the
air at fea †. But its ftructure is calculated for
the watery element, and, though it feldom or never
fwims, its four toes are connected by a fingle
fcalloped membrane. In this refpect, it ap-

" as they are dropt, and before they reach the water." *Hiftory of
the Buccaneers.*—" According to Oviedo, the Frigats wage the fame
" war againft the pelicans, when thefe repair to the Bay of Panama,
" to fifh for fardines." *Ray.*

   * Dampier.

   † Hence, in the Linnæan fyftem, the Frigat is denominated *Pelecanus Aquilus,* or Eagle Pelican.

proaches

proaches the cormorants, the boobies, and the pelicans, which may be regarded as perfect palmipeds. The bill of the Frigat is peculiarly calculated for rapine, since it terminates in a sharp hooked tip, and yet differs essentially from that of the birds of prey, being very long, the upper mandible somewhat concave, and the hook, placed quite at the point, seems to form a detached piece, as in the bill of the boobies, which it resembles by its sutures and by the want of external nostrils.

The Frigat is not larger than a hen, but its wings extend eight, ten, and even fourteen feet. This prodigious expansion enables it to perform its distant excursions, and transports it into the midst of the ocean, where it is often the only object between the sky and the water that gratifies the longing eyes of the mariner *; but this excessive length of wings has also its inconvenience; and, like the booby, the Frigat can hardly rise after it has alighted, so that when surprized in that situation it may be felled to the ground †. A cliff or the summit of a

tree,

---

* " We were accompanied with no bird in our route; a white
" booby or a Frigat appeared now and then at a great distance
" (between 15° and 20° south latitude)." *Cook.*

† " I went one of these last days to hunt Frigats on their islet at
" the extremity of Guadaloupe; we were three or four persons, and
" in less than two hours we took three or four hundred; we sur-
" prized the grown ones on the branches, or in their nest, and as they
" had great difficulty in taking wing, we had time to stun them with
" the

tree is required, and even then it cofts great effort to mount on wing *. We may fuppofe that all the palmated birds which perch have no object in view but to commence more eafily their flight; for that habit is not fuited to the ftructure of their feet, and it is only on elevated points that they can difplay their enormous wings and exert their pinions.

Hence the Frigats retire to fettle on the high cliffs or woody iflets, to breed undifturbed †. Dampier remarks, that they build their nefts on trees, in fequeftered fpots near the fea; they lay one or two eggs, which are white, with a carnation tinge, and having fmall dots of crimfon. The young ones are at firft covered with a light gray down; their feet are of the fame colour, and their bill is almoft white ‡ : but this colour afterwards changes, and the bill grows

---

" the blows of fticks." *Dutertre*.—" They leave their eggs with diffi-
" culty, and fuffer themfelves to be knocked down with fticks: I
" have often been witnefs and actor of this butchery." *M. de la
Borde*.

   * Dutertre.

   † " The fea rocks, and the little defert ifles, are the retreats of
" thefe birds; and in fuch fequeftered fpots they neftle." *Hift. Nat.
& Mor. des Antilles*.—" Thefe birds had very long poffeffed a little
" ifle in the extremity of Guadeloupe, to which all the Frigats of the
" neighbourhood came to repofe at night, and neftle in the feafon.
" It was called the *iflet of Frigats*, and ftill bears that name, though
" they have changed their retreat; for in the years 1643 and 1644,
" many perfons hunted them fo clofely, that they were obliged to
" forfake the iflet." *Dutertre*.

   ‡ Obfervation made by the Vifcount de Querhoënt at the ifland
of Afcenfion.

red

§

red or black, and bluifh in the middle; the fame
alteration takes place in the toes. The head is
pretty large, and flat above; the eyes are large,
black and brilliant, and encircled by a bluifh
fkin *. Under the throat of the adult male, there
is a large flefhy membrane of bright red, more
or lefs inflated or pendulous. No perfon has
diftinctly defcribed thefe parts; but if they
belonged exclufively to the male, they might
bear fome analogy to the caruncle of the turkey-
cock, which fwells and reddens, when the bird
is ftimulated by love or rage.

The Frigats are diftinguifhed afar at fea, not
only by the exceffive length of their wings, but
by the very forked fhape of their tail †. The
whole plumage is commonly black with a bluifh
glofs, at leaft that of the male ‡. Thofe which
are brown ‖, as the *little Frigat* figured by Edwards, feem to be females. Among the number of Frigats feen by the Vifcount de Querhoënt at the ifland of Afcenfion, and which were
all of the fame fize, fome appeared entirely
black, others of a deep black on the upper fur-

* Feuillee.
† The Portuguefe call the Frigat *Rabo Forcado,* on account of its
very forked tail.
‡ Ray.
‖ " The feathers of the back and of the wings are black, thick,
" and ftrong; thofe which cover the ftomach and thighs are more
" delicate, and not fo black. There are fome which have all the feathers
" brown on the back and on the wings, and gray under the belly; it
" is faid that the latter are females, or perhaps young ones." *Labat.*

face

face of the body, with the head and belly, white. The feathers on their neck are fo long, that the inhabitants of the South-Sea Iflands work them into bonnets *. They fet great value on the fat, or rather oil, extracted from thefe birds, on account of its fuppofed virtue in curing rheumatifms and torpors †.

This bird has, like the booby, the fpace round the eye naked; and alfo the nail of the mid-toe indented within. Thus the Frigats, though born the perfecutors of the boobies, are related to them by confanguinity: fad example in nature, of animals, which, like ourfelves, find often the moft inveterate foes among their kindred!

* Moft of the men at Eafter Ifland wore on their head a fillet of grafs, decorated with the long black feathers found on the neck of Frigats; others had enormous bonnets of the feathers of the brown gull. *Cook.*

† The oil or fat of thefe birds is a fovereign remedy in fciatic complaints, and for all others that originate from cold; it is efteemed a precious medicine in the Weft Indies. *Dutertre.*—The Buccaneers extract this oil, which they call *the oil of Frigats*, by boiling thefe birds in great cauldrons; it fells very dear in our iflands. *M. de la Borde*—The fat fhould be warmed, and rubbed well upon the part affected, in order to open the pores, and fpirit of wine fhould be mixed with it when the application is made: many people have received a complete cure, or at leaft great relief, from the remedy which I here mention on the credit of another, not having myfelf had an opportunity of putting it in practice. *Labat.*

[A] Specific character of the Frigat, *Pelecanus Aquilus:* " Its " tail is forked, its body black, its bill red, its orbits black."

The GULLS and the MEWS

LES GOELANDS *et* LES MOUETTES. *Buff*.

THESE two names, sometimes conjoined, sometimes separated, have hitherto served rather to confound than to discriminate the species comprehended in one of the most numerous families of the aquatic birds. Many naturalists have termed those *Gulls*, which others call *Mews*, and some have considered these two appellations as synonymous. But of all expressions in language, some traces must remain of their origin, or some marks of their differences: and I conceive that *Gull* and *Mew* correspond to the Latin words *larus* and *gavia*. I am persuaded also, that the Gulls properly include the larger, and the Mews the smaller species. Nay, we may discover vestiges of the same division among the Greeks; for the word κεπφος, which occurs in Aristotle, Aratus, and other authors, seems to denote a particular kind of Gulls. Suidas and the scholiast of Aristophanes render κεπφος by

---

* In Greek Λαρος and Κεπφος: in Eustathius Κηξ; and Lycophron gives the old ones the name Καυηξ, which seems to imitate their cry: in Latin *Larus* and *Gavia*: in German *Mew*: in Greenlandic *Akpa* or *Naviat*.

larus;

larus; and Gaza might have given the fame ver-
fion in his edition of Ariftotle *, had he not followed the conjecture of Pierius, that Virgil, in a
paffage of his Georgics, tranflated literally the
verfes of Aratus, and fubftituted *fulica* for the
Greek term. But if the *fulica* of the ancients
be the fame with our coot, the property afcribed
to it by the Roman poet, of playing on the
beach previous to a ftorm †, would be without
foundation ‡, fince that bird does not live on the
fea. The character which Ariftotle gives of his
κεπφος, that it fwallows the falt fpume, and is
caught by that bait, can apply only to a voracious bird, fuch as the Gull or the Mew. Aldrovandus accordingly concludes, after comparing thefe circumftances, that the λαρος in Ariftotle
is *generic*, and that the κεπφος is fpecific, or rather belongs to fome fubordinate fpecies of the
fame genus. But a remark which Turner has
made on the voice of thefe birds, feems to throw
us again into uncertainty: he conceives that the
word κεπφος is imitative of that of the Mew,
which ufually concludes its fhrill cries by a low,
fhort accent, or a fort of fneezing, *keph*; while

---

* Lib. ix. 135.

† ——————— *cumque marinæ*
In ficco ludunt fulicæ —— Virg. Georg. i. 362.

‡ The epithet which Cicero gives to the coot, in tranflating the
fame idea of Aratus, applies not to that bird, but agrees well with
the gull:

Cana fulix itidem fugiens è gurgite ponti,
Nuntiat horribiles clamans inftare procellas.

the

the gull terminates its fcream by a deeper tone, *cob*.

The Greek name κεπφος will correfpond then, in our divifion, to the Latin *gavia*, and will properly denote the inferior fpecies, or *the Mews*: while the appellation λαρος, or *larus*, will fignify the larger fpecies, or *the Gulls*. And to fix a term of comparifon in this fcale of magnitude, we fhall reckon all thofe birds *Gulls*, which exceed a duck in bulk, and meafure eighteen or twenty inches from the point of the bill to the end of the tail; and all under that dimenfion we fhall denominate *Mews*. It would thence follow, that the fixth fpecies, which Briffon calls the *firft Mew*, ought to be ranged with the Gulls, and that many of the Gulls in the Linnæan fyftem muft be claffed with the Mews: but before we defcend into the detail, we fhall exhibit the general characters and habits common to the whole genus.

All the Gulls and Mews are alike voracious and clamorous; they might be ftyled the vultures of the fea: they devour carrion of every kind which floats on the furface, or is caft on fhore. As cowardly as they are gluttonous, they attack only weak animals, and vent their fury on dead bodies. Their ignoble port, their importunate cries, their edged and hooked bill, prefent the hateful picture of birds fanguinary and bafely cruel. They fight rancoroufly together on the fcene of carnage; and even when they

§ are

are fhut up, and their ferocious humour is foured
by captivity, they wound each other without
apparent motive, and the firft from which blood
is drawn falls a victim to the reft ; for their fury
then rifes to a pitch, and they tear in pieces the
wretch which they had wounded without caufe *.
This excefs of cruelty is fcarce feen but in the
large fpecies ; but all of them, when at liberty,
continually watch an opportunity to fteal the
food or prey of their companions. Every thing
is acceptable to their voracity † : fifh, whether
frefh or putrid ; bloody flefh, recent or tainted ;
fhell-fifh, and even bones ; all digeft in their ftomach ‡. They fwallow the bait and the hook ;
they dart with fuch violence as to transfix themfelves on the point where the fifherman places
the herring or pilcher as a fnare. Nor is this
the only way to allure them ; Oppian afferts,
that if a board be painted with figures of fifh,
thefe birds will dafh againft it.—But ought not

* *Obfervation made by M. Baillon, of Montreuil-fur-mer.*

† " I have often given my Mews buzzards, ravens, new-born
" kittens, rabbits, and other dead animals ; they devoured them as
" greedily as they would do fifh : I have ftill two which can eafily
" fwallow ftares and fea-larks without plucking a feather ; their
" throat is a gulph which devours every thing." *Note communicated by M. Baillon.*

‡ " They difgorge fuch fubftances when they have plenty of
" other food ; but when they are pinched for want, the whole re-
" mains in their ftomach, and diffolves by the heat. Extreme
" voracity is not the only character in which thefe birds approach
" the vultures and the other ravenous birds ; the Mews fuffer hun-
" ger with equal patience : I faw one live nine days befide me,
" without tafting food." *Note of the fame obferver.*

thefe

thefe portraits to be as perfect as thofe of the grapes by Parrhafius?

Both the Gulls and the Mews have a long cutting bill, flat on the fides, with the point fortified and bent into a hook, and a protuberant corner at the lower mandible. Thefe characters are more apparent and decided in the Gulls, but yet occur in all the fpecies of Mews: by thefe they are diftinguifhed alfo from the terns, which have neither the hook on the upper mandible, nor the protuberance on the lower; not to mention that the largeft of the terns is inferior to the leaft of the Mews. The Mews have their tail not forked but entire: their leg, or rather their tarfus, is very high; and they would have the talleft legs of all the palmated birds, did not thofe of the flamingo, the avofet, and the long-fhank ftill exceed them, whofe ftructure is fo mifproportioned that they might be regarded as monftrous fpecies. All the Gulls and Mews have three toes connected by an entire membrane, and the hind toe detached, but very fmall: their head is large, and its carriage ungraceful, being funk almoft between the fhoulders, whether they walk or repofe. They run fwiftly on the beach, and fly ftill better above the waves: their long wings, which when clofed exceed the tail, and the quantity of feathers with which their body is clothed, make them very light *.

* We have a proverb, *You are as light as a Mew.* Martens.

They

They have also a very thick down *, which is of
a bluish colour, especially on the stomach. They
are hatched with that down, but the other feathers are late in growing; and they acquire not
completely their colours, to wit, the fine white
of the body, and the black or bluish gray on the
mantle, till after several moultings, and in their
third year. Oppian seems to have known this
progress of their colours; for he says that these
birds, as they grow old, become blue.

They keep in flocks on the sea-shores; some
running, some flying, and others alighting; the
beach and the downs seem quickened by their
numbers and their confused motions, and resound
incessantly with their noisy cries. In general,
no birds are more common on our coasts, and
they are found an hundred leagues at sea. They
frequent the islands and maritime countries of
every climate. Navigators meet with them in
all parts of the globe †. The larger species seem
attached

* Aldrovandus says, that in Holland the down of the Mew is much
used; but it would be hard to believe what he adds, that this down
heaves up at full moon, by a sympathetic concord with the swelling
of the tide.

† The Gulls are as common in Japan as in Europe. *Kæmpfer.*—
There are different kinds of them at the Cape of Good Hope, whose
cry is like that of the European Gulls. *The Viscount de Querhoënt.*
—As long as we were on this bank, which extends as far as Cape
Needles (off Madagascar) we saw Gulls. *Cook :*—he also saw Gulls
at Cape Froward, in the Straits of Magellan; at New Holland; at
New Zealand; near Statenland; in all the low islands of the south-
ern Archipelago; and many of the natives of Easter Island wore a

wooden

attached to the ſhores of the northern ſeas *. It
is reported that the Gulls of the Feroe iſlands are
ſo ſtrong and voracious, that they often tear the
lambs in pieces, and tranſport the fragments to
their neſts †. In the icy ocean, they often gather
in multitudes about the carcaſes of whales ‡ ;
and on theſe maſſes of corruption they fear not
infection. With ſuch repaſts they eaſily ſatiate
their rapacity, and procure ample proviſion for
the innate gluttony of their young. Theſe birds
ſtrew their eggs and neſts by thouſands. even on
the frozen lands of the two polar zones ‖ ; nor
do

wooden hoop decked with the white feathers of Gulls, which waved
in the air.—Clouds of Gulls produce in a great meaſure the dung
which covers the iſland of Iquique, and which is carried, under the
name of *guana*, into the valley of Arica. *Le Gentil.*—The Gull of
Louiſiana is like that of France. *Dupratz.*—A number of Gulls
and other birds came (at the Malouine iſlands) to hover on the
water, and darted upon the fiſh with extreme ſwiftneſs ; they led us
to diſcover the proper ſeaſon for catching ſardines ; if held a moment ſuſpended, they threw up that fiſh entire as it was juſt ſwal-
lowed : theſe birds lay round the pools, on green plants like the water-lily, a great number of excellent wholeſome eggs. *Bougain-*
ville.

* They abound on thoſe of Greenland to ſuch a degree, that
the wretched inhabitants of that frozen region have a peculiar word
to ſignify the hunting of this unpalatable game ; *akpalliarpok.*

† Forſter.

‡ Hiſt. Gen. des Voyages, *tom.* xix. 48.

‖ On the 5th of June we had already ſeen lumps of ice, which ſurprized us ſo much, that we took them at fiſſt for ſwans ... On the
11th, beyond the latitude of 75°, we landed on the iſland of *Baeren,*
where we found numbers of Gulls eggs. *Barentz.*—We advanced
as far as the iſland which Oliver Noorts had named *King's Iſland,*
(near the ſtraits of Le Maire) ; ſome ſailors who went aſhore, found
the

do they quit thofe regions in the gloom of winter, but feem attached to their native climates, and fcarcely affected by the change of temperature *. Ariftotle, who lived under a fky infinitely milder indeed, remarked that the Gulls and Mews never difappear, but remain the whole year in the places of their nativity.

The fame obfervation holds with refpect to France ; for many fpecies of this bird are feen on our coafts, both in fummer and in winter: on the weftern fhores they are called *mauves*, or *miaules*, and on the fouthern *gabians*. Every where they are noted for their voracity and their difagreeable importunate cries. Sometimes they keep on the low fhores, fometimes they retire into the cavities of the rocks, expecting the waves to caft out their prey ; often they attend the fifhers, to pick up the refufe and garbage: and this habit is doubtlefs the only ground of the affection towards man, which the ancients afcribed to thefe birds †. As their flefh is unfit for eating ‡, and their plumage of

no

the ground almoft entirely covered with eggs of a particular kind of Gull; one might reach forty-five nefts with his hand, without changing place, and each contained three or four eggs, rather larger than thofe of lapwings. *Le Maire* and *Schouten*.

* The birds which pafs in greateft numbers towards Hudfon's Bay in fpring, to breed in the north, and which return to the fouthern countries in autumn, are the ftorks, the geefe, the ducks, the teals, the plovers . . . but the Gulls fpend the winter in the country, amidft ice and fnow. *Hift. Gen. des Voyages, tom.* xv. 267.

† Oppian.

‡ We could not have tafted it without vomiting, if we had not previoufly expofed them in air, hanging by the claws, with their

heads

no value, they are neglected by the fowler, and
fuffered to approach without being fired upon*.

Curious to obferve by ourfelves the habits of
thefe birds, we fought to procure fome alive; and
M. Baillon, who is ever ready to oblige us, fent
a large Gull with a black mantle of the firft
fpecies, and a gray-mantled Gull of the fecond
fpecies. We have kept them fifteen months in
a garden, where we could obferve them at all
times. They fhowed at firft evident figns of
their malevolent temper, purfuing each other
continually, the large one never permitting the
fmall one to eat befide him. They lived on
foaked bread, the guts of game and of poultry,
and other offals from the kitchen, none of which
they ever refufed. They alfo gathered worms
and fnails in the garden, and could eafily feparate the fhells. They often went to bathe in a

---

heads downwards for feveral days, that oil or whales fat might drop
from their body, and that they might lofe their rank tafte. *Recueil des Voyages du Nord, tom.* ii. 89.

* The favages of the Antilles, however, eat thefe unfavoury
birds.—" There are," fays Father Dutertre, " numbers of fmall
" iflands fo full of them, that all the favages in paffing load their
" canoes with them; it is droll to fee how thefe people prepare
" them; they throw them entire, without gutting or plucking
" them, into the fire, and the burnt feathers form a cruft within
" which the bird is cooked. When they purpofe to eat, they re-
" move this cruft, then half open the bird. I know not how they
" preferve the carcafe from corruption, for I have feen fome that
" had been roafted eight days before, which is the more furpriz-
" ing, as in twelve hours moft kinds of flefh in thofe countries run
" into putrefaction."

fmall

ſmall baſon, and on coming out of the water they ſhook themſelves, clapped their wings, and then preened their plumage, like the geeſe and ducks. They roved at night, and were often ſeen walking out at ten or eleven o'clock. They do not, like moſt other birds, conceal their head under their wing when they ſleep; they only turn it behind, reſting the bill between the upper ſide of the wing and the back.

When a perſon tried to catch theſe birds, they bit and pecked with rancour: to maſter them without ſuffering injury, it was neceſſary to throw an handkerchief over their head. If one purſued them, they quickened their pace by ſpreading their wings. Uſually they walked ſlowly and ungracefully. Their ſloth was betrayed even in their rage; for when the largeſt purſued the other, he walked leiſurely, without ſeeming to care whether he ſhould overtake it; nor did the other ſhow any anxiety to eſcape, and when it reckoned itſelf to be at a ſufficient diſtance, it ſtopped; and it repeated this exertion as often as it was preſſed, ſo as to keep always beyond its enemy's reach, as if remoteneſs was ſufficient to deſtroy the antipathy.—Muſt not thus the weak ever retire for ſafety before the ſtrong? But unfortunately tyranny, in the hands of man, is an engine which extends as far as his thought!

Theſe birds appeared the whole winter to
forget

forget the ufe of their wings. They difcovered
no inclination to fly away: they were indeed
well fed, and their gluttonous appetite could not
torment them. But in the fpring they felt new
appetites, and fhowed other defires; they endeavoured to rife into the air, and would have
efcaped, had not feveral inches been clipt from
their wings: they could therefore only fpring
by jerks, or whirl on their feet with their wings
expanded. The paffion of love, which wakens
with the feafon, feemed to fupprefs the inftinct
of antipathy, and deftroy their mutual enmity:
for they feemed to fondle each other, and
though they did not confort, being of different
fpecies, they ate, flept, and refted together. But
their plaintive cries and reftlefs motions, fufficiently declared, that the fweeteft fentiment of
nature was provoked, not fatisfied.

THE BLACK BACKED GULL.

The BLACK-MANTLED GULL *.

FIRST SPECIES.

Larus Marinus. Linn. and Gmel.
Larus Niger. Briff.
Larus Maximus Varius. Ray, Will. and Klein.
The Great Black and White Gull. Alb. and Penn.
The Black-backed Gull. Stillingfleet and Latham.

THIS is the largeft of all the Gulls; it is two feet, and fometimes two feet and an half, in length: a great mantle of black or flaty blackifh covers its broad back; all the reft of its plumage is white: its bill is firm and ftrong, about three inches and a half long, yellowifh, with a red fpot on the projecting angle of the lower mandible: the eye-lid is orange-yellow; its feet, with their membrane, are flefh-coloured, whitifh and mealy.

The cry of this great Gull, which we kept a whole year, is a hoarfe found, *qua, qua, qua,* uttered in a raucous tone, and repeated very faft: but this is feldom heard, and when the bird is taken it vents a doleful and very fhrill fcream.

* In Swedifh *Homaoka:* in Danifh *Swart-bag, Blaa-maage:* in Norwegian *Hav-maafe:* in Lapponic *Gairo:* in Icelandic *Swart-bakur:* in Greenlandic *Naviarlurfoak.*

[A] Specific character of the *Larus Marinus:* " It is white, " its back black." Befides the northern parts of Europe, it is fpread over the whole of North America, and in the Carolinas it is, on account of its fober garb, called *the old wife.*

The GRAY-MANTLED GULL.

SECOND SPECIES.

Larus Glaucus. Gmel.
Larus Cinereus. Briff.
The Glaucous Gull. Penn. and Lath.

ASH-GRAY, fpread on the back and fhoulders, is the livery common to many fpecies of mews, and which diftinguifhes this Gull:
it is rather fmaller than the preceding, and except its gray mantle, and the black furrows on
the great quills of the wing, its plumage is entirely white. The eye is brilliant, and the iris
yellow as in the hawk : the feet are a livid
flefh-colour : the bill, which in young ones is
almoft blackifh, is pale yellow in adults : there
is a red fpot at the fwell of the lower mandible,
a character common to many fpecies of gulls
and mews. This bird flies from the preceding,
and dares not difpute with it about its prey.
But it revenges itfelf on the mews, which are its
inferiors in ftrength : it pillages them, purfues
them, and wages continual war with them. It
frequents much, in the months of November
and December, the coafts of Normandy and
Picardy ; where it is called the *large miaulard*

and

THE GLAUCOUS GULL.

and *blue-mantle*, as the appellation *black-mantle*
is beſtowed on the firſt ſpecies. This bird had
ſeveral diſtinct cries which it uttered in the
garden, where it lived with the preceding ; and
the firſt and moſt frequent of theſe ſeemed to
be the two ſyllables *qui-ou*, which began like a
whiſtle, ſhort and ſharp, and cloſed with a
drawling tone, lower and ſofter. It was repeated only at intervals, and to produce it the
bird was obliged to make an effort, and to extend its neck and incline its head. The ſecond
cry, which was never vented but when purſued,
or held cloſe, and which was therefore an expreſſion of fear or anger, may be imitated by the
ſyllable *tia, tia*, whiſtled and repeated very faſt.
—We may obſerve, by the way, that of all animals the cries of anger or fear are ſharper and
ſhorter than the uſual accents.—About ſpring
this bird aſſumes a very ſhrill, piercing voice,
which might be denoted by the word *quieute* or
pieute, ſometimes ſhort and rapidly pronounced,
ſometimes produced on the final *eute*, with diftinct intervals, like the ſighs of a perſon in dif-
treſs. In either caſe, this cry ſeems to be the
plaintive expreſſion inſpired by unſatisfied love.

[A] Specific character of the *Larus Glaucus* : " It is white;
" its back and wings hoary ; its wing-quills tipt with white; its
" bill bright yellow, with the corner ſaffron."

The BROWN GULL.

THIRD SPECIES.

Larus Catarrhactes. Linn. and Gmel.
Larus Fuscus. Briff.
Catarractes. Gefner, Sibbald, and Will.
Catarracta. Aldrov. Johnft. and Charleton.
The Skua Gull. Penn. and Lath.

THIS Gull has its plumage of an uniform dufky brown over the whole body, except the belly, which is ftriped acrofs with brown on a gray ground, and the great quills of the wing, which are black. It is fomewhat fmaller than the preceding, its length from the bill to the extremity of the tail being only a foot and eight inches, and an inch lefs to the nails, which are fharp and ftout. Ray obferves, that the whole habit of the body of this bird befpeaks rapine and carnage; and fuch indeed is the bafe and cruel afpect of all the fanguinary tribes of gulls. It is to this fpecies that naturalifts have generally referred the *Catarractes* of Ariftotle *, which, as the word imports †, defcends with rapidity to feize its prey; which agrees well

* Hift. Animal. *lib.* ix. 12.

† From Καλα downwards, and Ρɛω to flow: hence alfo the word *cataract.*

with

with what Willughby reports of the great gull,
that it dafhes its head againft the board on which
the fifhermen fix the bait. The *Catarractes* of
Ariftotle is undoubtedly a marine bird, fince,
according to the philofopher, it drinks fea-
water *. The Brown Gull in fact occurs on
the ocean, and the fpecies appears to be fettled
in the high latitudes on both fides of the equator: it is common in the Feroe iflands, and on
the coafts of Scotland †. It feems even more
diffufed on the fhores of the South Sea; and is
probably the bird which our navigators denominate *the cobler (cordonnier ‡)* without hinting at
the

* Nothing furely is more falfe than what Oppian fays, that the
Catarractes contents itfelf with dropping its eggs upon the fea-
weeds, and leaving them to be foftered by the wind; unlefs it be
what he adds, that about the time they are to hatch, the male and
female take each in their claws an egg which is to produce a chick
of their own fex, and let them fall repeatedly on the fea, till by
this exercife the young birds burft from their prifon.

† *Catarractes Nofter.* Sibbald.

‡ According to the notes which the Vifcount de Querhoënt has
had the goodnefs to communicate, the cobler occurred in his route,
not only near the Cape of Good Hope, but in higher or lower
latitudes in the open fea. This obferver feems to diftinguifh a
greater and a leffer fpecies of thefe birds, as will appear from the
following extract:

" I believe that the inhabitants of the water live more friendly
" and focially than thofe of the land, though of very different fpe-
" cies and fizes: they alight pretty near each other without any
" miftruft; they hunt in company, and I never but once faw a
" fight between a frigat and the leffer kind of cobler; it lafted
" pretty long in the air, and each defended itfelf with its wings
" and its bill: the cobler, though incomparably weaker, eladed by
" its agility the formidable blows of its antagonift, and did not

the reason of that appellation. The English
found numbers of them at Port-Egmont, in the
Falkland or Malouine islands, and have therefore stiled them *the Port Egmont hens* *. We
cannot do better than transcribe the account
given in the second voyage of the celebrated
Captain Cook.

" yield; it was vanquished, when a petrel which chanced to be
" near repaired to the scene, and, by passing and re-passing seve
" ral times between the combatants, effected their separation; the
" cobler through gratitude followed its deliverer, and came with it
" round the ship."

* On the 24th of February, in latitude 44° 40', on the coasts
of New Zealand, Mr. Banks, being in the boat, killed two Port-
Egmont hens, exactly like those which we had found in great
numbers on the island of Faro, and which were the first that we
saw on this coast, though we had met with several a few days before
we discovered land. *Cook.*—In 50° 14' latitude south, and 95° 18'
longitude west, as many birds were flying about the ship, we took
the opportunity of the calm to kill some of them ; one was of the
species which we have so often mentioned under the name of the
Port Egmont hen, a sort of gull nearly of the size of a raven, and
of a deep brown plumage, except below each wing, where there
were some white feathers; the rest of the birds were albatrosses and
sheer-waters. *Idem.*—On the islands near Statenland, we counted
among the sea-birds the *Port Egmont hens. Idem.*—The birds which
we met with in Christmas Sound, near Terra del Fuego, were
geese, ducks, sea-pies, shags, and that kind of gull so often mentioned in this Journal by the name of *Port Egmont hen. Idem.*—
There were also (at New Georgia) albatrosses, common gulls, and
that kind which I call *Port Egmont hen. Idem.*—In the latitude of
54° 4', we also saw a Port Egmont hen, and some weed. Navigators have generally looked upon all these to be certain signs of the
vicinity of land ; I cannot, however, support this opinion. At
this time we knew of no land, nor is it probable there is any nearer
than New Holland or Van Diemen's land, from which we were
distant two hundred and sixty leagues. *Idem.*

" In

" In the evening, being in the latitude of
" 64° 12′ fouth, longitude 40° 15′ eaft, a bird
" called by us in my former voyage Port Eg-
" mont hen, (on account of the great plenty of
" them in Falkland Ifles) came hovering feveral
" times over the fhip, and then left us in the
" direction of N. E. They are a fhort thick
" bird, about the fize of a large crow, of a dark
" brown or chocolate colour, with a whitifh
" ftreak under each wing in the fhape of a half-
" moon. I have been told that thefe birds are
" found in great plenty at the Feroe Iflands
" north of Scotland, and that they never go far
" from land. Certain it is I never before faw them
" above forty leagues off; but I do not remem-
" ber ever feeing fewer than two together;
" whereas here was but one, which, with the
" iflands of ice, may have come a good way
" from land *."

* (To this account Mr. Forfter adds, that he recognized it to
be the great northern gull, *Larus Catarractes,* common in the high
latitudes in both hemifpheres; that a few days after they faw another of the fame kind, which rofe to a great height above their
heads, which they regarded as a novelty, the birds of that climate keeping near the furface of the water.)

[A] Specific character of the *Larus Catarrhactes:* " It is
" grayifh; the quills of its wings and tail are white at the bafe; its
" tail fomewhat equal."

The VARIEGATED GULL; *or*, the GRISARD.

FOURTH SPECIES.

Larus Catarrhactes. Linn. and Gmel.
Larus Nævius. Gmel.
Larus Marinus, var. 2. Lath. Ind.
Larus Varius, five Skua. Briff.
Larus Major. Aldrovandus.
Larus Cinereus Major. Charleton.
Larus Grifeus Maximus. Klein.
The Wagel of the Cornifh. Ray.
The Sea Eagle. Sibbald's Hift. Fife.
The Skua Gull. Penn.
The Wagel Gull. * Lath. Syn.

THE plumage of this Gull is broken, and ftreaked with brown gray on a white ground; the great quills of the wing are black- ifh; the bill black, thick and ftout, and four inches long. This Gull is one of the largeft, its alar extent being five feet, which meafure was taken from an individual fent alive from *Montreuil-fur-Mer,* by M. Baillon. This bird lived a long time in a court-yard, where it killed its companion by fighting: it fhowed the fordid

---

* In Holland it is called *Mallemucke*; in the Feroe iflands *Skua*; in Norway *Skue, Kav-orre.*

familiarity

THE WAGEL GULL.

familiarity of a voracious animal, which hunger
only attaches to the hand that feeds it. It
fwallowed flat fifh almoft as broad as its body;
and with equal avidity it devoured raw flefh,
and even fmall animals entire, fuch as moles,
rats, and birds *. A gull of this kind, which
Anderfon received from Greenland, attacked
fmall animals, and fiercely defended itfelf with
its bill againft dogs and cats, and took pleafure
in biting their tails. On fhowing it a white
handkerchief, it was fure to fcream with a
piercing tone, as if that recalled to its memory
fome foe which it dreaded at fea.

All the Grifards are, according to the obfervations of M. Baillon, of a dirty and dark gray
when young; but after the firft moult the
tint grows more dilute, the belly and the neck
whiten firft, and, in three moults, the plumage
is entirely waved and freckled with gray and
white, fuch as we have defcribed it. The
white afterwards continues to gain ground, and
the aged birds are entirely hoary. If the plumage, therefore, were the fole foundation of
diftinction, we fhould admit an unneceffary
number of fpecies, fince nature varies to fuch
degree the colours with the years.

In this, as in all the other gulls and mews,

* Whence probably the fable of Oviedo *(Hift. Ind. Occid.* lib.
xiv. 18.) has been applied to the Grifard, of a bird which has one
foot webbed for fwimming, and the other armed with talons for
feizing its prey.

the

the female appears rather fmaller than the male. Belon remarks, that it is not common in the Mediterranean, that it feldom occurs in our interior provinces *, but is numerous on our weftern coafts. It roams very far to fea, fince we are affured of its being found in Madagafcar † But the congenial region of this fpecies feems to be the North.   Thefe birds are the firft which the veffels meet in approaching Greenland ‡ ; and they conftantly attend thofe employed in the whale-fifhery, following them even amidft the ice. When a whale is killed, they alight in myriads on the floating carcafe, and tear it on all fides § : and though the fifhers labour to drive them away, by ftriking with poles and oars, they can hardly, without felling them, make them quit their hold ‖.   This fenfelefs obftinacy has occafioned the Dutch name *Mallemucke* or *ftupid beaft* ¶.   " Thefe foolifh, fordid

* M. Lottinger pretends that he has feen fome of thefe birds on the great pools of Lorraine, in the fifhing feafon ; and M. Hermann fpeaks of a Grifard killed in the neighbourhood of Strafburg.

† *Note communicated by Dr. Mauduit.*

‡ Klein.

§ The herrings afford abundant fubfiftence to thefe legions of birds : Zorgdrager fays, that he faw a quantity of herring bones about the nefts of the water-fowl on the rocks of Greenland. *Peche de la Baleine, partie* ii. 7.

‖ Memoirs of the Academy of Stockholm.

¶ From *mall*, which fignifies *fottifh* or *ftupid* ; and *mocke*, which in old German means *beaft* or *animal*.

" birds,

" birds, quarrel and fight," fays Martens,
" fnatching from each other their morfels,
" though the large carcafes on which they feed
" might abundantly fatiate their voracity."

Belon perceived fome analogy between the
head of the Variegated Gull and that of the
eagle : but, in its fordid groveling habits, it
much more refembles the vultures. Its ftout,
hardy conftitution renders it capable of bearing
the moft inclement weather ; and mariners have
remarked, that it cares little for ftorms at fea.
It is well clothed with feathers, which feem to
form the chief part of the bulk of its very lean
body. But we are not certain if all thefe birds
be conftantly lean ; for the one which we
chanced to fee, had a hook fticking in its
palate, and grown over with callous flefh, which
muft have hindered it from fwallowing eafily.

According to Anderfon, it has an air-bag under its fkin, like that of the pelican *. This
naturalift obferves, that his Greenland *Mallemucke* differs in fome refpects from that of
Spitzbergen, defcribed by Martens. We muft
notice that Martens himfelf feems to join, under this name *Mallemucke*, two birds, which at

* He adds fome anatomical details : " Each lobe of the lungs
" is formed like a feparate lung in fhape of a purfe ; the cryftalline
" of the eye is fpherical like that of fifhes ; the heart has only one
" chamber ; the bill is perforated with four noftrils, two difclofed,
" and two concealed under the feathers at the root of the bill."
Hift. Nat. d'Iflande & de Groënland.

other

other times he difcriminates; and the fecond,
or that of Spitzbergen, from the ftructure of
its bill, *articulated with feveral pieces*, and having *tubular noftrils*, and alfo *its croaking like frogs*,
appears to be a petrel, rather than a gull.—To
this fpecies we may alfo refer a race or variety,
larger than the common, and whofe plumage is
rather waved than fpotted or ftriped: it is defcribed by Lidbeck, and occurs in the gulf of
Bothnia; fome individuals are eight or ten
inches longer and broader than the common
kinds of *Grifards*.

[A] Specific character of the *Larus Nævius* : " It is white; its
" back cinereous; its tail-quills tipt with black,"

The BROWN-GRAY-MANTLED GULL ; *or*, the BURGOMASTER.

FIFTH SPECIES.

Larus Glaucus. Linn. and Gmel.
Larus Cinereus. Briff.
Burgermeifter. Martens and Klein.
The Glaucous Gull. Penn. and Lath. *

THE Dutch who frequent the northern feas
on the whale fifhing, are conftantly at-

* In Sweden it is called *Maos :* in Norway *Krykie :* in Lapland
Skierro : and in Greenland *Tatarrok.*

tended

tended by clouds of mews and gulls. They
have fought to diftinguifh them by names fig-
nificative or imitative, *mallemucke, kirmew, ratfhet,
kutgegef.* The prefent they have ftiled the *Burgomafter,* becaufe, by its ftature and grave de-
portment, it would feem to prefide as a magif-
trate among thefe diforderly and voracious
tribes *. It is indeed a fpecies of the firft magnitude, and as large as the black-mantled gull.
Its back is brown gray, and alfo the quills of
the wing, of which fome are tipt with white,
others with black, the reft of the plumage
white: the eye-lid is edged with red or yellow: the bill is yellow, with the lower angle
very protuberant, and of a bright red; which
Martens well expreffes, by faying, that it feems
to have a red cherry on the bill. It was probably from overfight that this traveller neglects
to reckon the hind-toe, which is indeed very
fmall: for it is evidently the fame with the *herring gull* † of the Englifh. In the northern feas,
thefe birds live on the carcafes of large fifh.
" When a whale is dragged after a veffel," fays

* There are prodigious numbers of thefe aquatic birds in Greenland: we find there all the fpecies defcribed by Martens, in his
Voyage to Spitzberg, and many others which he does not mention.

† *Larus Fufcus.* Linn. and Gmel.
Larus Grifeus. ⎫ Briff.
Gavia Grifea. ⎭
Larus Cinereus Maximus. Ray, Will. and Sloane.
Specific character: " It is white; its back brown."

Martens,

Martens, " they flock about it, and fteal large
" pieces of the blubber. They may then be
" eafily killed; but it is impoffible to gain its
" neft, which it places on the fummit and in
" the clefts of the higheft rocks. The *Burgo-*
" *mafter*," he adds, " intimidates the *mallemucke*,
" which, however ftout it be, fubmits to be
" beaten and pecked, without attempting to re-
" taliate. When the Burgomafter flies, its
" white tail fpreads like a fan. Its cry refem-
" bles that of a raven: it purfues the young
" *lumbs*, and often hovers about the fea-horfes,
" whofe dung it feems to fwallow."

According to Willughby, the eggs of this
gull are whitifh, fprinkled with fome blackifh
fpots, and as large as hen's eggs. Father Feuillée mentions a bird on the coafts of Chili
and Peru, which, by its figure, its colours, and
its voracity, refembles this northern gull, but
which is probably fmaller; for that travelled
naturalift fays, that its eggs are only fomewhat
larger than thofe of the partridge. He fubjoins, that he found its ftomach entirely filled
with the feathers of certain fmall birds of the
coafts of the South Sea, which the natives call
tocoquito.

The GRAY and WHITE-MANTLED GULL.

SIXTH SPECIES.

Larus Fuscus. Linn. and Gmel.
Gavia Grisea. Briff.
The Herring Gull. Penn. and Lath.

IT is probable that this Gull, defcribed by Father Feuillée, and which is nearly as large as the gray-mantled Gull, is only a fhade or variety of that fpecies, or of fome of the preceding, at a certain age. Its figure and ftructure feem to lead to that inference. Its mantle, fays the Jefuit, is gray mixed with white, and fo is the upper furface of the neck, of which the fore fide is light gray, and all its *livery:* the quills of the tail are dull red lead colour, and the top of the head is gray. He adds, as a fingular property, that the inner toe has only two joints, the middle one three, and the outer four, which is therefore the longeft; but this ftructure, the moft favourable for fwimming, fince the broadeft part of the fole has thus the greateft compafs of motion, occurs in a great number of aquatic birds, and even among the waders. We have obferved particularly in the jacana, the fultana, and the water hen, that the outer toe contained four *phalanges,* the middle one three, and the inner only two.

8

The WHITE MEW.

FIRST SPECIES.

Larus Eburneus. Gmelin and Phipps.
The Ivory Gull. Penn. and Lath.

FROM what we have faid of the wagels, which whiten with age, we might fuppofe that this is only an old one; but it is much fmaller than that gull; its bill is neither fo large, nor fo ftrong, and its plumage is pure white, without any tinge or fpot of gray. It exceeds not fifteen inches from the end of the bill to that of the tail. It is defcribed in Captain Phipps's * voyage to the north : he obferves very properly, that this fpecies has not been delineated in the Linnæan fyftem, and that the bird called *raths-herr*, or fenator, by Martens, perfectly refembles it, except in the feet, to which Martens attributes only three toes. But if we fuppofe the fourth toe, which is very mi-

---

* " The whole bird is fnowy and fpotlefs; its bill lead-coloured; " its orbits faffron, afh-leaden ; its nails black; its hind-toe jointed " and nailed; its wings longer than its tail; its tail equal, and longer " than its feet. The whole length of the bird, from the tip of the " bill to the end of the tail, is fixteen inches ; the diftance between " the tips of the fpread wings, thirty-feven inches; the bill two " inches."

nute,

nute, to have efcaped the obfervation of that navigator, it would exactly correfpond to his *raths-
herr*. Its whitenefs furpaffes that of fnow; and
the ftately pace of the bird on the ice has procured it the appellation of *raths-herr*, or fenator.
Its voice is low and ftrong; and whereas the
little mews or *kirmews* feem to call *kir* or *kair*, the
Senator founds *kar*. It is ufually folitary, unlefs
fome prey collects a certain number of them.
Martens faw them alight on the carcafes of fea-
horfes, and devour their dung.

[A] Specific character of the *Larus Eburneus* : " It is all white,
" its orbits faffron, its bill and feet lead-coloured."

The SPOTTED MEW; *or*, the KUTGEGHEF.

SECOND SPECIES.

Larus Riga. *Larus Tridactylus.*	} Gmel.
Gavia Cinerea Nævia. *Gavia Cinerea.*	} Briff.
Larus Cinereus Pifcator.	Klein.

The Tarrock of the Cornifh.
The Kittiwake of the Scotch.

" WHILE we were cutting up the whale-
" blubber," fays Martens, " a number
" of thefe birds came fcreaming about our fhip,
                                        " and

" and feemed to pronounce *kutgeghef*." That
found refembles, indeed, the fort of fneezing,
keph, *keph*, which feveral captive gulls utter, and
from which we conjectured the Greek name κεπφοc
was derived. This bird exceeds not in bulk the
white mew; it is only fifteen inches long: the
plumage confifts of a fine white ground on the
fore fide of the body, and gray on the mantle,
marked with fome ftreaks of the fame gray that
form a kind of half-collar on the upper fide of
the neck; it is diftinguifhed alfo by fpots of
white and black mixed on the coverts of the
wing, with varieties, however, which we fhall
mention. The hind toe, which is very fmall in
all the mews, is fcarce perceptible in this one,
as Belon and Ray obferve. And hence Martens
fays, that it has only three toes : he adds, that it
always flies rapidly againft the wind, however
violent this blows; but that it is perpetually
purfued and harraffed by the bird *ftrundt-jager* *,
and conftrained to void its excrements, which
the latter greedily fwallows. In a fubfequent
article †, we fhall find that this depraved tafte
has been erroneoufly imputed to the *ftrundt-
jager*.

This Spotted Mew occurs not only in the feas
of the north, it alfo inhabits the coafts of Eng-
land ‡ and Scotland ||. Belon, who met with it

* i. e. *Dung-hunter*.
† See the article of the *Dung-bird*.
‡ Ray. || Sibbald.

in

in Greece, fays, that he recognized it merely
from the name *laros*, which it ftill bears in that
country : and Martens, after having obferved it
at Spitzbergen, found it again in the Spanifh
feas, fomewhat different, indeed, yet ftill diftin-
guifhable ; whence he very judicioufly infers,
that animals of the fame fpecies in diftant countries muft ever receive impreffions from the climates.   So great is the difference in the prefent
cafe, that this fpecies has been fplit into two:
the *cinereous mew* of Briffon, and his *cinereous
fpotted mew*, are unqueftionably the fame, as a
comparifon of the figures will evince.   And
what completely eftablifhes our pofition, is a
feries of fubjects, which exhibits a gradual progrefs of the black and white of the wing, from
the mottled colours to the fimple gray.   The
gray half-collar on the top of the neck is common to all the individuals of this fpecies.

Flocks of thefe Mews appeared fuddenly near
Semur in Auxois, in the month of February 1775.
They were very eafily killed, and were found
dead or half-ftarved with hunger in the meadows, the fields, and the brinks of rivulets.   On
opening them, their ftomach was found to contain fome fragments of fifhes, and their inteftines a blackifh jelly.   Thefe birds were not
known in the country; their appearance lafted
only a fortnight; they were brought by a ftrong
fouth wind, which blew all that time *.

* *Obfervation communicated by M. de Montbeillard.*

[A] Specific

[A] Specific character of the Kittiwake, *Larus Riga:* " It is
" white, its back hoary, its tail-quills entirely white, its feet have
" three toes.''—Specific character of the Tarrock, *Larus Tridacty-*
lus: " It is whitish, its back somewhat hoary, its tail-quills, except
" the outermost, tipt with black ; its feet have three toes." But
Mr. Latham very properly classes them under the same species, the
Tarrock being only the young bird.

The GREAT CINEREOUS MEW; *or*, the BLUE-FOOTED MEW.

THIRD SPECIES.

Larus Canus. Linn. and Gmel.
Gavia Cinerea. Briss.
Galedor, Crocala, Galetra. Aldrov.
Gabiano Minore. Zinn.
Larus Rostro Nigro. Klein.
Larus Cinereus Minor. Will. Ray, and Sibb.
The Common Sea-Mall, or *Mew.* Will. and Ray.
The White Web-footed Gull. Albin.
The Common Gull. Penn. and Lath.

THE bluish colour of the feet and bill, constant in this species, ought to distinguish it
from the others, which have the feet generally
of flesh-colour, more or less vermilion or livid.
It is sixteen or seventeen inches long, from the
point of the bill to the end of the tail ; its mantle is light cinereous ; several of the wing-quills
are furrowed with black ; all the rest of its
plumage is snowy white.

Willughby

Willughby reckons this the moſt common ſpecies in England. It is called the *grand emiaulle* * on the coaſts of Picardy. M. Baillon has made the following obſervations on the different ſhades of colours that its plumage aſſumes in the ſucceſſive moultings: in the firſt year the quills of the wings are blackiſh; and not till the ſecond moulting do they acquire the diſtinct black and the white ſpots with which they are variegated: no young Mew has a white tail, the end is always black or gray; at the ſame age the head and the upper ſide of the neck are marked with ſome ſpots, which are by degrees obliterated, and give place to pure white: the bill and the feet gain not their full colour till two years old.

To theſe general obſervations, very important as they are to ſtop the unneceſſary multiplication of ſpecies from individual varieties, M. Baillon adds ſome on the particular nature of the Blue-footed Mew. It is more difficult to tame than the reſt, yet it ſeems not ſo wild in the ſtate of liberty: it fights leſs, and is not ſo voracious as moſt of the others; but it is not ſo ſprightly as the little cinereous Mew. When kept in a garden, it ſought earth-worms: if offered ſmall birds, it would not touch them till they were half-torn; which ſhews that it is not

---

* i. e. *The Great Mew,* from *miauler,* to ſquall like a cat; whence the Engliſh verb *to mewl,* and the names of theſe birds *Mall* and *Mew.—T.*

fo carnivorous as the gulls. And as it is not fo lively or cheerful as the little Mews, which remain to be defcribed, it feems, by its fize and its inftincts, to hold the middle rank between them both.

[A] Specific character of the Common Sea-Mew, *Larus Canus:* " It is white, its back hoary." It is the moſt numerous of all the gulls, at leaſt in Great Britain. It breeds on the ledges of the cliffs that overhang the fea.

———

The LITTLE CINEREOUS MEW.

FOURTH SPECIES.

Larus Cinerarius. Linn. and Gmel.
Gavia Cinerea Minor. Briff.
Larus Cinereus Primus. Johnſt.
The Red-legged Gull. Lath.

ITs inferior fize, and the different colour of its legs, diftinguiſh this Mew from the preceding, which it refembles exactly in its colours. It has the fame light cinereous and bluiſh on its mantle, the fame black fcallops fpotted with white on the great quills of the wing, and laftly, the fame fnowy white over all the plumage, except a black fpeckle, which appears
pears

pears conftantly on the fides of the neck behind
the eye. The livery of the young ones confifts
of brown fpots on the coverts of the wing. In
the aged, the feathers of the belly have a flight
tint of rofe colour; and it is not till the fecond
or third year, that the legs and the bill degenerate
from a fine red into a livid complexion.

This and the laughing Mew are the two leaft
of the whole family. They exceed not the
bulk of a large pigeon, and their body is much
thinner: they are thirteen or fourteen inches
long. They are very handfome, clean, and active; lefs vicious than the large fpecies, yet
more lively. They eat many infects, and during
fummer they make a thoufand evolutions in the
air after beetles and flies. They take fuch
quantities of thofe, that their ftomachs are filled
up to their bill. They follow the rife of the
tide * in the rivers, and fpread fome leagues over
the land, groping in the marfhes for worms and
leeches, and return in the evening to the fea.
M. Baillon, who made the obfervations, adds,
that they might eafily be made to inhabit gardens, where they would feed on infects, fmall
lizards, and other reptiles. Yet they may be
kept on foaked bread, but muft always have
much water, becaufe they every inftant wafh
their bill and feet. They are very clamorous,

* Sometimes they advance very far: M. Baillon faw one on the
Loire, above fifty leagues from its mouth.

efpecially

efpecially when young; and on the coaft of Picardy they are called the *petites miaulles*, (the
little Mews). It feems that the name *tattaret*
has alfo been applied on account of their cry *.
They appear to be the fame with the gray gulls
mentioned in the Portuguefe relations of India,
under the denomination *garaios*, and which navigators meet with in numbers on the paffage
from Madagafcar to the Maldives †. To fome
fimilar fpecies alfo we muft refer the bird called
tambilagan in Luçon, and which is a gray Mew
of fmall fize, according to the fhort defcription
given by Camel in his account of the Philippine
birds, inferted in the Philofophical Tranfactions.

* " The *tattaret* is a little common gull; it derives its name from
" its cry. It is the fmalleft, but the handfomeft of the birds of this
" clafs: it would be enti ely white, were not its back azure. The
" *tattarets* build in flocks, on the fummits of the moft craggy rocks;
" and if a perfon approaches them, they begin to fly with fhrill
" cries, as if they would frighten people away with the hideous
" noife." *Hift. Gen. des Voyages, tom.* xix. *p.* 47.

† On this track there are always feen numbers of birds, fuch as
gray gulls, which the Portuguefe call *garaios* . . . Thefe gulls come
to alight on the veffels, and fuffer themfelves to be caught by the
hand, without fearing the fight of men, as having never experienced
them: they had the fame fate with the flying-fifh, which they hunt
on thefe feas, and which, being purfued at once by the birds and the
fifhes, often throw themfelves on board the veffels. *Voyages qui
ont fervi a l'Etabliffement de la Compagnie des Indes Orientales; Amfterdam,* 1702, *tom.* i. *p.* 277.

[A] Specific character of the *Larus Cinerarius :* " It is white,
" its back hoary, with a brown fpot behind the eyes."

The LAUGHING MEW.

FIFTH SPECIES.

Larus Ridibundus. } Linn. and Gmel.
Larus Atricilla. }

Gavia Ridibunda Phœnicopos. } Briff.
Gavia Ridibunda. }

Coppbus Turnerl. Gefn.

Larus Cinereus. Ray, Will. and Scop.

The Pewit, or *Black Cap, Sea Crow,* or *Mire Crow.* Will.

The Pewit. Plott's Staffordfhire.

The Brown-headed Gull. Albin.

Baltner's Great Afh-coloured Sea Mew. Will.

The Pewit Gull. Penn.

The Black-headed Gull, Lath. *

The Laughing Gull. Catefby, Penn. and Lath.

THE cry of this little Mew bears fome refemblance to a hoarfe laugh; and hence
its epithet. It is fomewhat larger than a pigeon; but, like all the mews, its body appears
much more bulky than in reality. The quantity of fine feathers with which it is clothed,
makes it very light; hence it flies almoft continually over the water; and during its fhort intervals
on land, it is extremely buftling and noify, particularly in the breeding feafon, when thefe
birds are moft collected †. It lays fix olive

* In German *Groffer See-Swalle* (greater fea-fwallow) and *Grauer
Fifcher* (gray fifher): in Polifh *Rybitw Popielafly Wiekfy*: in Mexican *Pipixcan*.

† Ray.

2C 3   eggs

eggs fpotted with black; the young ones are good food, and, according to the British Zoology, they are taken in great numbers in the counties of Effex and Stafford.

Some of thefe Laughing Mews fettle on the rivers and even the pools of inland countries *; and they feem to frequent the feas of both continents. Catefby found them at the Bahama Iflands: Fernandez defcribes them under the Mexican name *pipixcan:* and, like all the other mews, they abound moft in the northern countries. Martens, who obferved them in Spitzbergen, and calls them *kirmews,* fays, that they lay on whitifh mofs, in which the eggs can hardly be diftinguifhed, being alfo dirty white, or greenifh dotted with black : they are as large as thofe of pigeons, but very fharp at the end; the yolk is red, and the albugineous liquor is bluifh. Martens fays, that he ate of them, and found them very good, tafting like the eggs of lapwings. The parents dart boldly on the perfon who dares to rob their neft, and with loud cries they endeavour to drive him off by ftriking with their bills. The firft fyllable of their name *kirmew,* is expreffive of their notes, according to the fame traveller; who remarks, however, that their voice differs in the various regions which they inhabit, the polar tracts, the

---

* Kramer and Schwenckfeld. Thefe birds are feen on the Thames, near Gravefend, according to Albin.

coafts

coafts of Scotland, thofe of Ireland, and the German Ocean. He afferts that, in general, a difference may be perceived in the cries of animals of the fame fpecies, refulting from the influence of climate; and this diverfity may indeed obtain, efpecially in birds, for the tones of animals are the expreffions of their moft ufual fenfations; and the feathered race are delicately fenfible to the variations of the atmofphere, and to the impreffions of temperature.

Martens obferves likewife, that the Mews of Spitzbergen have finer and more hairy feathers than thofe of our feas. This difference, too, arifes from the climate. Another, which feems to be derived from the age of the individual, confifts in the colour of the bill and the feet: in fome thefe are red, in others black. But what proves that this difference does not conftitute two diftinct fpecies is, that the intermediate fhade occurs in feveral fubjects; fome having the bill red, and the legs only reddifh, and others having the bill red at the tip only, and the reft black. Thus we admit but one fpecies of Laughing Mew; the difference which led Briffon to make a fubdivifion, lying entirely in the colour of the bill and of the legs. In the female, the front and throat are marked with white, whereas the whole head of the male is covered with a black cap: the great quills of the wing are alfo partly this colour: the mantle is bluifh cinereous, and the reft of the body white.

[A] Specific

[A] Specific character of the Black-headed Gull, *Larus Ridi-bundus:* " It is whitish, its head blackish, its bill and feet red."—Specific character of the Laughing Gull, *Larus Atricilla:* " It is " somewhat hoary, its head blackish, its bill red, its feet black."

We shall transcribe Dr. Plott's account of the manner of catching the Pewits last century in Staffordshire. After relating some mar-vellous stories respecting their attachment to the lord of the manor, he thus proceeds : " Being of the migratory kind, their first ap-" pearance is not till the latter end of February, and then in number " scarce above six, which come as harbingers to the rest, to see whe-" ther the hafts or islands in the pools (upon which they build " their nests) be prepared for them ; but these never so much as " lighten, but fly over the pool, scarce staying an hour: about " the 6th of March following, there comes a pretty considerable " flight of an hundred or more, and then they alight on the hafts, " and stay all day, but are gone again at night. About our Lady-" day, or sooner in a forward spring, they come to stay for good, " otherwise not till the beginning of April, when they build their " nests, which they make not of sticks, but heath and rushes, " making them but shallow, and laying generally but four eggs, " three and five more rarely, which are about the bigness of a small " hen egg. The hafts or islands are prepared for them between Mi-" chaelmas and Christmas, by cutting down the reeds and rushes, " and putting them aside in the nooks and corners of the hafts, and " in the valleys, to make them level ; for should they be permitted " to rot on the islands, the pewits would not endure them.

" After three weeks sitting the young ones are hatched, and about " a month after they are almost ready to fly, which usually happens " on the 3d of June, when the proprietor of the pool orders them to " be driven and catched, the gentry coming from all parts to see the " sport: the manner is thus—They pitch a rabbet-net on the bank-" side, in the most convenient place over-against the hafts, the net " in the middle being about ten yards from the side, but close at " the ends in the manner of a bow ; then six or seven men wade into " the pool, beyond the pewits, over-against the net, with long staves, " and drive them from the hafts, whence they all swim to the bank-" side, and landing, run like lapwings into the net, where people are " standing ready to take them up, and put them into two pens made " within the bow of the net, which are built round, about three " yards diameter, and a yard high, or somewhat better, with small " stakes driven into the ground in a circle, and interwoven with " broom and other raddles."—(This description is illustrated by an engraving).

" In

" In which manner there have been taken in one morning fifty
" dozens at a driving, which, at five shillings a dozen (the ancient
" price of them) comes to £. 12. 10 s.: but at several drifts that have
" been anciently made in the same morning, there have been as
" many taken as have sold for £. 30; so that some years the profit of
" them has amounted to £. 50 or £. 60, beside what the generous
" proprietor usually presents his relations and the nobility and gen-
" try of the county withal, which he constantly does in a plentiful
" manner, sending them to their houses in crates alive ; so that feed-
" ing them with livers and other entrails of beasts, they may kill them
" at what distance of time they please, according as occasions pre-
" sent themselves, they being accounted a good dish at the most plen-
" tiful tables.

" But they commonly appoint three days of driving them, within
" fourteen days, or thereabouts, of the 2d or 3d of June ; which, while
" they are doing, some have observed a certain old one that seems to
" be somewhat more concerned than the rest, being clamorous, and
" striking down upon the heads of the men; which has given ground
" of suspicion that they have some government among them, and that
" this is their prince that is so much concerned for its subjects. And
" it is further observed, that when there is great plenty of them, the
" Lent-corn of the country is much the better, and so the cow-pas-
" tures too, by reason they pick up all the worms and the fern-flies,
" which, though bred in the fern, yet nip and feed on the young corn
" and grass, and hinder their growth." (The pools of Staffordshire,
which the pewits frequented, were Pewit Pool, in the parish of
Norbury, and Sebben Pool, in the parish of High Offley). *Plott's
Natural History of Staffordshire, pp.* 232 and 233.

The WINTER-MEW.

SIXTH SPECIES.

Larus Hybernus. Gmel.
Larus Canus, var. Lath. Ind.
Gabia Hyberna. Briff.
Larus Fufcus, feu Hybernus. Ray, &c.
Guaca-guacu. Marcgrave.
In Cambridge-fhire, Coddy-Moddy. Will.

WE fufpect that this bird is no other than
our fpotted mew *(kittiwake)* which vifits
the inland parts of England during winter. We
make this conjecture, becaufe its bulk is the
fame, and its plumage fimilar, only brown
where the other is gray : and it is well known
that thefe birds when young have a darker caft,
not to mention how eafily the fhades may be
confounded in a defcription or drawing. If
that of the Britifh Zoology had been better, we
could have fpoken with more confidence. However, this Mew lives in winter on earth-worms,
and when its ftomach is overloaded, it difgorges
them half-digefted; which is the origin of the
ftar-fhot or *ftar-gelly.*

*　　*　　*　　*　　*

[A] Specific character of the Winter-Mew, *Larus Hybernus*: " It is white; its top, the back of its head, and the sides of its neck, " spotted; its back cinereous; the first of its wing-quills black; its " tail marked with a black bar near the tip."—There is a passage in Morton's History of Northamptonshire, which, as it throws light on the œconomy of the Winter-Mew, and indeed on the nature of birds in general, deserves a place in this work.

" I shall here," says the intelligent author, " set down my remarks " upon that *gelatinous* body called *star-gelly, star-shot,* or *star-fall'n*; " so named because vulgarly believed to fall from a star, or to be the " recrement of the meteor which is called the falling or shooting star, " or rather the meteor itself shot down to the earth. It is generally a " clear, almost sky-coloured, tremulous, viscid, or tenacious gelly, but " in water glib and slippery. To the hand it is as cold as we gene- " rally say a frog is; and is apt to strike a chillness into it. 'Tis " found in lumps or masses of a certain size. The largest I have seen " was about the bigness of a goose egg. Those that are found fresh " and new, are generally entire, very little, if at all, broken or dif- " persed: some of the masses are curled or convoluted somewhat like " the *ileum,* or as a garden-snail appears when the shell is broken off " from it. In consistence and colour, it pretty much resembles boiled " starch, or rather a solution of *gum tragacanth.* I set some of this " gelly on the fire in a silver vessel: it did not dissolve, as does cold " boiled starch when it is set again upon the fire; but became thick " and viscous; the more fluid or watery part of it having gone off " in vapour. I let it stand till the humidity was all evaporated. To " the bottom of the vessel adhered certain skins and vessels, like those " of animal bodies. The inside of the porringer upon this operation " exhibited a glistering like that of isinglass; and there proceeded " from it a kind of greasy smell. Another experiment I made of the " gelly broken and put into spring-water, which I set to seeth upon " the fire, but not much of it dissolved. A piece of muslin being " dipt into it so managed as with starch, was stiffened as with that. " The gloss that it gave to the muslin might be rubbed off. The " gelly being put into water, some of it subsided; some did not. In " the pieces that floated I observed several bubbles, which I thought " were formed by the water intruded into and retained in the pores " of it.

" It is sometimes found with patches of a dirty yellow in it; " sometimes with black specks resembling little pellets of grumous " blood. Breaking one of the lumps, I discovered and took out of " it several pieces of tough skin, and of long tenacious string-like bo-
"dies.

" dies.   Macerating another of thefe maffes that I might clear the
" ftrings, as I called them, from the fofter or more pulpy part of the
" gelly, I found them branched and diftended through the whole
" mafs. The fmaller branches in fize and figure like capillary blood-
" veffels of a blackifh red. With thefe were broad pieces of a very
" thin film or membrane. The mafs, though a large one of the fort,
" might be drawn up and fufpended by thefe ftrings.   The gelly,
" when it has been kept for fome time, becomes putrid, and has a
" ftrong fmell, like that of a dead carcafe in a putrid ftate. Chickens
" will eat it.

" This ftar-fhot, as it is called, is very rarely to be met with on
" the tillage-lands. At Oxendon, and in moft other places, we find
" it chiefly in the lower and moifter ground, particularly on the
" ledges of fod, upon the fides of trenches in meadows, upon and near
" the banks of brooks or pools, on ant-hills and mould-banks now
" and then in great plenty. It occurs fometimes in dry, barren, and
" heathy ground. It does not appear for any number of years fuc-
" ceffively in one and the fame place. At Pisford, in January 1702,
" was fhown me a mafs of ftar-gelly lying upon a dead hedge that
" I am well affured had its firft appearance there. The like has been
" found upon a flat board at the top of a cherry-ladder, in Mr.
" Courtman's garden at Thorpe. The laft in as compact a mafs as
" thofe upon the ground are ufually found in.

" 'Tis chiefly feen in mifty mornings, and in wet weather, in au-
" tumn, in winter, and early in the fpring ; feldom or never any that
" is frefh and new in the time of froft, or in the fummer months.

" In 1700, there was no ftar-gelly to be found about Oxendon,
" till a wet week in the end of February, when the fhepherds brought
" me about thirty feveral lumps or maffes of it.

" Applied to the running heels of horfes, it has been found to be
" of ufe ; as alfo for pafting paper to pafteboard, glafs, and the like.

" As to the origin of this body, it has in many particulars a near
" analogy with animal fubftances ; as the defcription of it plainly
" fhows.   And by feveral other circumftances that are not men-
" tioned, it appears to me to be only the difgorging or cafting of
" birds of three or four forts ; of thofe fort of fowl in particular
" that at certain feafons do feed very plentifully upon earth-worms
" and the like ; and efpecially of the *fea-mew*, and the *winter-mew*
" or *coddy-moddy*, birds of the gull kind, which are very ravenous.
" The *coddy-moddies* come up into this country in great numbers at
" the time this ftar-fhot is generally found, viz. in the autumn and
" the winter months : frequent thofe very places where it ufually

" occurs,

" occurs, viz. moiſt meadows and the banks of brooks, more rarely
" ploughed lands ; and greedily devour earth-worms, which in
" thoſe places, and about that time of the year, are very nume-
" rous. They generally come up the vallies, where our brooks
" and rivers run, very early in the morning, even before the ſhep-
" herds or any body elſe are abroad in the fields ; eſpecially in foggy
" mornings and before a ſtorm, in ſuch ſeaſons and in ſuch weather
" as in a particular manner invite the earth-worms out of their
" holes and receſſes up upon the ſurface : and the birds return
" again to ſeaward. In the month of September, 1708, I ſaw a
" *coddy-moddy* ſhot down to the ground, that on her fall upon the
" ground, when almoſt half-dead, diſgorged a heap of half-digeſted
" earth-worms, much reſembling the gelly called ſtar-ſhot.

" In fine, having compared the notes or marks of that diſgorge
" or caſting with thoſe of the gelly, called ſtar-ſhot, I found them
" ſo much alike, I am ſatisfied the latter is for the main the ſame
" origin as is the former. Some of theſe maſſes I take to be diſ-
" gorged by herons and bitterns after having fed upon frogs, which
" they ſometimes do. Sir William Craven once ſhot a bittern by
" one of Winwick pools, which after great heaving and working
" of her breaſt, at length diſcharged a quantity of this ſort of gelly.
" The worthy Mr. Thomas Clerke, of Watford, aſſures me, that
" he has ſeen a maſs of ſtar-gelly, wherein appeared the head and
" other parts of a frog almoſt diſſolved into a gelly, like to that
" which encompaſſed it. Having kept a parcel of frogs ſpawn
" ſome time by me, it had a ſmell very like that of corrupted ſtar-
" ſhot. Others of them, it is not unlikely, are diſgorged by crows,
" when they feed over-abundantly on earth-worms. The carrion
" crow will likewiſe feed upon frogs and toads too, pecking them
" into pieces, and ſo devouring them ; whereas the herons, &c.
" ſwallow them whole. The gelly upon the dead hedge, and on the
" cherry-ladder, in the inſtances above-mentioned, I am apt to
" think came from crows or rooks. . . .

" 'Tis uſual with birds, the more ravenous ſort eſpecially, to caſt
" up what is uneaſy and burthenſome to their ſtomachs. This is
" well known to thoſe who are converſant and experienced in the
" buſineſs of ordering and managing of birds, eſpecially hawks.
" We may reaſonably ſuppoſe, that all other birds that have a
" membranous ſtomach, and voracious appetite, do the ſame upon
" any the like occaſion ; theſe in particular I have now mention-
" ed, namely, herons and bitterns, which have a membranous ſto-
" mach, as have all the carnivorous birds : and the Winter-Mew,

" and

AFTER this enumeration of the species of Gulls and Mews well described and distinctly known, we shall mention a few others, which might probably be ranged with the preceding, if their indications were more complete.

1. That which Brisson calls *the little gray mew*, saying, that it is *equal in bulk to the great cinereous mew*; and which seems to differ from that species, or from that of the gray-mantled bill, because it has white mixed with gray on the back.

2. That great Sea-Mew, mentioned by Anderson, which preys on an excellent fish, called in Iceland *runmagen*. He tells us, that this bird carries it ashore, and eats only the liver; and that the peasants instruct the children to run up to the Mew as soon as it alights, and snatch from it the fish.

3. The bird killed by Mr. Banks, in the latitude of 1° 7' north, and longitude 28° 50', and which he terms the *black-footed gull*, or *larus crepidatus*. The excrements were of a bright red,

" and the carrion crow, whose stomachs are not furnished with such
" thick muscles, as are those of the granivorous birds. These, the
" *Winter-Mew* particularly, having glutted and overcharged their
" stomachs with earth-worms, or the like cold and viscid food, they
" cast it up again not duly dissolved; then especially when the
" earth-worms, &c. are a new or more uncommon sort of food to
" them. We have a parallel instance in some sorts of fishes, accord-
" ing to the curious and judicious Mr. Ray, who takes the *Ballæ*
" *Marinæ*, which are little round lumps (some of them as big as
" tennis-balls) of *festucæ* amassed together, to be cast out of fishes
" stomachs."

3   approaching

approaching that of the liquor contained in the *helix*, which floats on the fea ; which renders it probable that the bird feeds on that fhell-fifh.

4. The Mew called by the inhabitants of the ifland of Luçon *taringting*, and which, from the character of vivacity afcribed to it, and its habit of running fwiftly on the fhore, may be either the little gray mew or the laughing mew.

5. The Mew of the lake of Mexico, called by the inhabitants *acuicuitzcatl*, and of which Fernandez fays nothing more.

6. Laftly, a Gull obferved by the Vifcount de Querhoënt, in the road off the Cape of Good Hope, and which, from the account he obligingly communicated, muft be a fort of the black-mantle, but its legs, inftead of red, are fea-green.

The LABBE, or DUNG-BIRD.

Larus Crepidatus. Gmel.
Catarracta-Cepphus. Will. and Ray.
Stercorarius. Briff.
Strund-jager, of the North.
The Black-toed Gull. Penn. and Lath.

CONSIDERING its fize and figure, this bird might be ranged with the mews. But if it be of that family, it has loft all fraternal affection ; for it is the avowed and eternal perfecutor of its kindred, and particularly of the kittiwake. It keeps a fteady eye on them, and when it perceives them betake to flight, it purfues without intermiffion. The people of the north report that its object is to obtain the excrements from thofe unhappy little mews; and they have, for that reafon, called it *Strundjager*, to which *Stercorarius* is fynonymous. Moft probably, however, this bird does not devour the dung, but only the fifh which the kittiwakes drop from their bill or difgorge * : efpecially

---

* Some naturalifts have alledged, that certain fpecies of gulls purfue others for their excrements. I have done all that was in my power to afcertain this fact, which I was always averfe to believe. I have frequently repaired to the fea-fhore to make obfervations,

THE BLACK TOED GULL.

efpecially as it catches fifh itfelf, and alfo eats whale's blubber; and amidft the abundance of nourifhment with which the fea fupplies its in-habitants, it would be very ftrange if the Labbe was reduced to fwallow what all the reft re-ject.

No perfon has better defcribed thefe birds than Ghifter, in the Memoirs of the Academy of Stockholm. " The flight of the Labbe," fays he, " is fwift and poifed, like that of the " gofhawk. The ftrongeft wind cannot hinder " it from catching in the air the fmall fifh " thrown to it by the fifhermen. When they

vations, and have difcovered what give rife to the fable. It is this:

The gulls maintain with each other a perpetual conteft about their carnage, at leaft the great and middle fpecies: when one comes out of the water with a fifh in its bill, the firft which per-ceives it fhoots down to fnatch the prey; and if the fortunate plun-derer haftens not to fwallow the capture, it will be purfued in its turn by others ftill ftronger, which ftrike it violently with their bill; it cannot avoid them but by efcaping, or by repelling its ene-my; and whether that the fifh incumbers its flight, or that it is over-come with fear, or fenfible that the fifh is the fole motive of the purfuit, it quickly throws it up; the other, which fees it drop, catches it dextroufly before it reaches the water, and feldom miffes to receive it.

The fifh appears always white in the air, becaufe it reflects the light, and feems, by reafon of the celerity of the flight, to drop be-hind the gull which vomits it. Thefe two circumftances have de-ceived obfervers.

I have verified the fame fact in my garden; I chafed fome large gulls fhouting after them; they ran and difgorged the fifh which they had juft fwallowed; I threw it to them, and they caught it in the air, with as much alertnefs as dogs. *Note communicated by M. Baillon, of Montreuil-fur-mer.*

" call *lab, lab,* it immediately repairs to receive
" the fish, whether raw or dreffed, or the other
" food which is offered to it. It alfo takes the
" herrings out of the buffes, and if they are falt-
" ed, it wafhes them before fwallowing. One
" can hardly approach it or fire upon it, un-
" lefs fome bait be thrown. But the fifhermen
" are kind to it, as it is an almoft infallible
" fign of a herring-fhoal ; and when the Labbe
" does not appear, their fuccefs is fmall. This
" bird is almoft always on the fea; generally
" two or three appear together, and very fel-
" dom five or fix. When it cannot find pro-
" vifion at fea, it comes to the beach and at-
" tacks the mews, which fcream on its appear-
" ance : but it rufhes on them, and overtaking
" them, it alights on their back, and obliges
" them to caft up the fifh which they had juft
" fwallowed. This bird, as well as the mews,
" lays its eggs on the rocks ; the male is
" blacker and rather larger than the female *."

Though it is the Long-tailed Labbe to which
thefe obfervations feem chiefly to apply, we conceive that they relate alfo to the fpecies now
under confideration, whofe tail is fafhioned fo
that the two middle feathers are the longeft,
but do not much exceed the others. Its bulk
is nearly equal to that of the little mew, and its
colour is dun cinereous, waved with gray-

---

* Colleftion Academique, *partie etrangere, tom.* xi. *p.* 51.

ifh :

ifh * : the wings are very large, and the legs are formed as in the mews, only not quite fo ftrong; the toes are fhorter. But the bill differs more from that of thefe birds, the end of the upper mandible being armed with a nail or hook that appears added; a charaƈter in which the bill of the Labbe refembles that of the petrels, though the noftrils are not tubular.

The Labbe has, in the carriage of its head, fomething of the bird of prey, and its predatory life belies not its appearance. It walks with its body ereƈt, and fcreams very loud: it feems, fays Martens, to pronounce *i-ja* or *johan*, when heard at a diftance, and its voice refounds. Their mode of life neceffarily difperfes them; and that navigator fays, that they are rarely found together: he adds, that the fpecies feemed not to be numerous, and that he met with few about Spitzbergen. The ftormy winds of the month of November, 1779, drove two of thefe birds upon the coafts of Picardy: they were fent to us by M. Baillon, and from them we have made the preceding defcription.

* This colour is lighter below the body; and fometimes, according to Martens, the belly is white.

[A] Specific charaƈter of the *Larus Crepidatus:* " Its two mid-" dle tail quills are longer than the reft." It is found in the north-ern parts of Europe and America, and even on the Atlantic. It weighs eleven ounces; its length is fifteen inches, and its alar extent thirty-nine. Linnæus fays, that it lays two eggs, which are pale ferruginous, fpotted with black.

The LONG-TAILED LABBE.

Larus Parasiticus. Linn. and Gmel.
Stercorarius Longicaudus. Briff.
Plautus Stercorarius. Klein.
Catarracta Parasitica. Brun.
The Arctic Bird. Edw.
The Arctic Gull. Penn. and Lath. *

THE production of the two middle feathers
of the tail in two detached and diverging
fhafts, characterifes this fpecies, which is of the
fame bulk with the preceding. It has a black
cap on the head; its neck is white, and all the
reft of its plumage gray: fometimes the two
long feathers of the tail are black. This bird
was fent to us from Siberia, and we think that
it is the fame fpecies with that found by Gmelin
in the plains of Mangafea, near the banks of the
river Jenifea. It occurs likewife in Norway,
and even in Finmark and Angermania: and
Edwards received it from Hudfon's Bay, where,
he obferves, the Englifh, no doubt on account
of its hoftilities againft the mew, call it the
Man-of-war Bird, a name beforehand applied,

* In Denmark it is called *Strondt-jager*, or *Schyt-valk* (dung-
bird): in Sweden *Swart-laffe*; and in Angermania *Labben*: in
Norway *Kyuffwa* or *Tjufva*.

and

and with better reafon, to the frigat.   That author adds, that, from the length of its wings and
the weaknefs of its legs, he fhould have judged
that this bird lived more flying at fea than
walking on land : yet, he remarks, the feet are
as rough as a file, and proper to cling to the
flippery backs of large fifhes.   Edwards enter-
tains the fame opinion with us, that the Labbe,
by the form of its bill, is intermediate between
the mews and petrels.

Briffon reckons a third fpecies of Labbe, the
Stercorarius Striatus * ; but, as it is founded on
Edwards' defcription, who regarded it as the
female of the Long-tailed Labbe, we cannot
adopt it.   We alfo are of opinion, that it is
only a variety from age or fex; and we even
fufpect that our firft fpecies might perhaps admit the fame arrangement.   In that cafe, we
fhould have only one kind of Labbe, of which
the adult or male would be that with two long
feathers in the tail, and the female would be that
reprefented by Edwards ; the mantle deep afh-
brown on the wings and tail, with the fore fide
of the body of a dirty white gray ; the thighs,
the lower belly, and the rump, croffed with
blackifh and brown lines.

* " Above brown; the feathers edged at the tip with tawny;
" below dirty white, ftriped tranfverfely with brown; its head
" brown; its throat whitifh brown; its tail-quills white at their
" origin, and deep brown the reft of their length." *Briffon.*

[A] Specific character of the *Larus Parafiticus:* " Its two middle tail-quills are very long."

The ANHINGA.

Plotus Anhinga. Linn. and Gmel.
The White-bellied Darter. Lath.

IF regularity of form in animals, and fymme-
try of proportion, ftrike us as graceful and
beautiful, and if the rank which we affign them
correfponds to the feelings they excite; nature
knows not fuch diftinctions. She loves them
becaufe they are the children of her creation;
and her attachment requires no other plea. She
cherifhes alike in the defert the elegant *gazel* *
and the fhapelefs camel; the pretty mufk † and
the gigantic *giraff* ‡ : fhe launches into the air
at once the magnificent eagle and the hideous
vulture; and fhe conceals under the earth and in
the waters generations innumerable of infects,
fafhioned in every fantaftic fhape. All varieties
of figure and ftructure fhe admits, provided they
are fuited to the fubfiftence and propagation of
the kind. The *mantes* live under the form of a
leaf: the fea-urchins are imprifoned within a
fpherical fhell; the vital juices filter and circulate

---

* *Antilope-Dorcas.* Linn. *The Barbarian Antelope.* Penn.
† *Mofchus Pygmæus.* Linn. *The Guinea Mufk*, Penn.
‡ *Camelopardalis-Giraffa.* Linn. *The Camelopard.* Penn.

through

THE WHITE-BELLIED AHNGO.

through the branches of the *afterias*. The head
of the *zygena* is flattened into a hammer; and the
whole body of the moon-fiſh is rolled into
a ſpiny globe. And do not a thouſand other
productions of figures equally ſtrange demon-
ſtrate, that the univerſal mother has aimed at
diffuſing animation, and of extending it to all
poſſible forms? Not content with varying the
traces and ſhades of the original pictures, does
not ſhe ſeem ſolicitous to draw communicating
lines from each genus to all the others; and
thus, from her rudeſt ſketches to her moſt
finiſhed performances, all are connected and
interwoven? Thus we have ſeen that the
oſtrich, the caſſowary, and the dodo, by the
ſhortneſs of their wings, the weight of their
body, and the largeneſs of the bones of their legs,
form the ſhade between the quadrupeds and the
birds: the penguins are a-kin to the fiſh: and
the Anhinga, the ſubject of this article, exhibits
a reptile grafted on the body of a bird. Its ex-
ceſſively long and ſlender neck, and its ſmall
cylindrical head, rolled out like a ſpindle, of the
ſame girth with the neck, and drawn out into a
long ſharp bill, reſemble both the figure and the
motion of a ſnake, whether the bird nimbly extends its head to fly from the tops of trees, or
unfolds it and darts it into the water to pierce
the fiſhes.

Theſe ſingular analogies have equally ſtruck
all who have obſerved the Anhinga in its na-

tive

tive country * (Brazil and Guiana) ; they ftrike us even in the dried fpecimens of our cabinets. The plumage of the neck and head does not alter its flender fhape ; for it is a clofe down, fhaven like velvet : the eyes are of a brilliant black, with the iris golden, and encircled by a naked fkin : the bill is jagged at the tip with fmall indentings turned backwards. The body is fcarcely feven inches long, and the neck alone meafures double.

The extreme length of the neck is not the only difproportion that ftrikes us in the figure of the Anhinga. Its large and broad tail, compofed of twelve fpread feathers, differs no lefs from the fhort round fhape which obtains in moft of the fwimming birds : yet the Anhinga fwims, and even dives, only holding its head out of the water, in which it plunges entirely on the leaft fufpicion of danger : for it is very wild, and can never be furprized on land. It keeps conftantly on the water, or perched on the talleft trees, by the fides of rivers and in overflowed favannas. It builds its neft on thefe trees, and repairs among them to pafs the night. Yet it is entirely palmated, its four toes being connected by a fingle piece of membrane, and the nail of the middle one ferrated within. Thefe coincidences of ftructure and habits feem to indicate an affinity between the Anhinga and the cormorant

* Marcgrave.—Barrere.

and

and boobies; but its fmall cylindrical head, and
its bill drawn out to a point, without any hook,
diftinguifh it from thefe two kinds of birds.——
The fkin of the Anhinga is very thick, and the
flefh commonly fat, but has a difagreeable oily
tafte: Marcgrave found it to be no better than
that of the gull, which is furely very bad.

None of the three Anhingas figured in our
Planches Enluminees exactly refembles that defcribed by Marcgrave. Nᵘ 960 has, like that
naturalift's, the upper fide of the back dotted,
the end of the tail fringed with gray, and the
reft of a fhining black: but all the body is
black, the head and neck are not gray, and the
breaft is of a filvery white. Nº 959 has not
the tail fringed *. Yet we think that thefe
two birds, which were brought from Cayenne,
are really of the fame fpecies with the Brazilian
Anhinga, defcribed by Marcgrave; the differences of colours not exceeding what, in the
plumage of the water birds efpecially, might
refult from age or fex. Marcgrave remarks
too, that the nails of his Anhinga were reflected and very fharp, and that it ufes them to
catch fifh; that its wings are large, and reach,

* *Plotus Melanogafter.* Gmel.
The Black-bellied Anhinga. Penn.
The Black-bellied Darter. Lath.

Specific character: " Its head is fmooth; its belly black."

when

when clofed, to the middle of its long tail. He
feems, however, to over-rate its bulk in comparing it to the duck. The Anhinga which we
know, may be about thirty inches, or even more,
from the tip of the bill to the end of the tail:
but this large tail and its long neck occupy the
largeft fhare of this meafure, and its body does
not appear to exceed that of a morillon.

[A] Specific charaƈter of the *Plotus Anhinga* : " Its head is
" fmooth ; its belly white."

The RUFOUS ANHINGA.

Anhinga Melanogafter, var. 3. Gmel.

WE have feen that the Anhinga is a native
of South America, and, notwithftanding
the poffibility that fuch a bird, furnifhed with
long wings, might traverfe the ocean, like the
cormorants and the boobies, I fhould have re-
ftriƈted it to thofe countries ; nor would the
denomination merely of Senegal Anhinga have
altered my opinion, had not a note of Adanfon, accompanying a fpecimen, affured us, that
a fpecies of Anhinga inhabits the coaft of
Africa, where the people of the country call

5   it

it *kandar*. This Senegal Anhinga differs from those of Cayenne, because its neck, and the upper fide of its wings, are of a rufous fulvous, marked by pencils on a dark brown ground, the reft of the plumage being black. Its figure, its port, and its bulk, are exactly the fame as in the American Anhingas.

The SHEARBILL.

Le Bec-en-Ciseaux. *Buff*.

Rhynchops Nigra. Linn. and Gmel.
Rygchopfalia, Briff.
The Cutwater. Catefby.
The Black Skimmer. Penn. and Lath.

THE mode of life, the habits, and œconomy of animals, are not fo free as might be fuppofed. Their actions refult not from inclination and choice, but are the neceffary effects of their peculiar organization and ftructure. Nor do they feek ever to infringe or evade the law of their conftitution: the eagle never abandons his rocks, or the heron her fhores: the one fhoots down from the aerial regions, to plunder or murder the lamb, founding his prefcriptive right on his ftrength, his armour, and his habitual rapine; the other, ftanding in the mire, patiently expects the glimpfe of its fugitive prey. The woodpecker never forfakes the trees, round which he is appointed to creep. The fnipe muft for ever remain in its marfhes; the lark in its furrows, and the warbler in its groves. All the granivorous birds feek the inhabited countries, and attend on the progrefs of cultivation. While
thofe

THE BLACK SKIMMER.

thofe which prefer wild fruits and berries, perpetually fly before us, and cherifh the wilds, and
forefts, and mountains : there, remote from the
dwellings of man, they obey the injunctions of
nature. She retains the hazel grous under the
thick fhade of pines ; the folitary blackbird under his rock ; the oriole in the forefts, which re-
found with its notes ; while the buftard feeks its
fubfiftence on the dry commons, and the rail in
the wet meadows. Such are the eternal, immu-
table decrees of nature, as permanent as their
forms : thefe great poffeffions fhe never refigns,
and on thefe we vainly hope to encroach. And
are we not continually reminded of the weaknefs
of our empire ? She obliges us even to receive
troublefome and noxious fpecies : the rats make
a lodgment in our houfes, the martins in our
windows, the fparrows in our roofs; and when fhe
conducts the ftork to the top of our old ruinous
towers, already the habitation of the mournful
family of nocturnal birds, does fhe not haften
to refume the poffeffions which we have ufurped
for a time, but which the filent lapfe of ages
will infallibly reftore to her ?

Thus the numerous and diverfified fpecies of
birds, led by inftinct, and confined by their
wants to the different diftricts of nature, have
apportioned among themfelves the air, the earth,
and the water. Each holds its place, and enjoys its little domain, and the means of fubfiftence, which the extent or defect of its faculties
will

will augment or abridge. And as all the poffible gradations in the fcale of exiftence muft be filled up, fome fpecies, confined to a fingle mode of fupport, cannot vary the ufe of thofe imperfect inftruments which nature has beftowed on them. Thus the fpoonbill feems formed for gathering fhell-fifh : the fmall flexible ftrap and the reflected arch of the avofet's bill, reduce it to live on fifh-fpawn : the oyfter-catcher has an ax-fhaped bill, calculated for opening the fhells : and the crofsbill could not fubfift, were it not dextrous in plucking the fcales from the fir-cones. Laftly, the *Shearbill* could neither eat fidewife, nor gather food, nor peck forwards; its bill confifting of two pieces extremely unequar, the lower mandible, being long and extended difproportionately, projects far beyond the upper, into which it falls like a razor into its haft *. To catch its prey with this awkward and defective inftrument, the bird is obliged to fly, fkimming the furface, and with its lower mandible cutting the water. By this neceffary and laborious exertion, the only one it can perform, it fhovels up the fifh, and earns its fubfiftence †. Hence fome obfervers have called it *cutwater :* the name *Shearbill (bec-en-cifeau)* is derived from

* Ray.

† They feed on fmall fifh, which they catch flying where the water is fhallow; they keep their lower mandible almoft always in the water, and when they feel a fifh they clofe both mandibles, which may be termed the blades. *Memoirs on the Natural Hiftory of Guiana, communicated by M. de la Borde, king's phyfician at Cryenne.*

the

the ftructure of its bill; the lower mandible being hollowed out by a channel, and furnifhed
with two fharp ledges, receives the upper one,
which is flattened like a blade.

The point of the bill is black, the part next
the head is red, and fo are the feet, which have
the fame ftructure as thofe of the gulls. The
Shearbill is nearly equal to the little cinereous
mew: the whole upper furface of the body, the
fore fide of the neck, and the front, are white:
it has alfo a white ftreak on the wing, fome of
whofe quills, and alfo the lateral ones of the tail,
are partly white : all the reft of the plumage is
black, or blackifh brown: in fome fubjects
it is fimply brown, which appears to indicate a variety from age * ; for, according to
Catefby, the male and female are of the fame
colour.

Thefe birds are found on the coafts of Carolina and of Guiana; on the latter they are numerous, and appear in flocks, almoft always on
wing, and only alight in the mire.   Though
their wings are very long, their flight is remarked
to be flow † : if it were fwift, they could not diftinguifh and raife their prey, as they rufhed along

* *Rynchops Fulva.* Linn.
Specific character: " It is fulvous, its bill black." However,
Gmelin reckons it only a variety.

† *Memoirs communicated by M. de la Borde.*

the

the furface of the water.   According to the obfervations of M. de la Borde, they come in the
rainy feafons to neftle on the iflets, and particularly the *Grand Connetable*, near the fhores of
Guiana.

The fpecies feems peculiar to the American
feas; nor can we extend it to the Eaft Indies,
becaufe Ray's continuator mentions a drawing
fent from Madras, but which was perhaps made
elfewhere.   We are alfo of opinion, that the
fheerwater of the South Sea, fo often mentioned
by Captain Cook, is not the fame with the
Shearbill of Cayenne, though they have received
the fame name: for, befides the immenfe difference between the hot climate of Cayenne and
the pinching colds of the South Seas, it appears, from two paffages of his narrative, that
his fheerwaters were petrels *, and that they
occur in the higher latitudes and even on
the frozen iflands with the albatroffes and
penguins †.

---

* " We now began to fee fome of that fort of petrels fo well
" known to failors by the name of fheerwaters, latitude 58° 10′ S.
" longitude 150° 54′ E." *Second Voyage*, vol. i. p. 45.—" We had
" another opportunity of examining two different albatroffes, and a
" large black fpecies of fheerwater, *Procellaria Æquinoxialis :* we
" had now been nine weeks out of fight of land." *Idem.*

The fheerwater is the fame with the puffin, which will afterwards
be defcribed, and which is in fact a fpecies of petrel.

† " We were in the midft of the ice (in 61° 51 S. and 95° E.);
" we had but few birds about us; they were albatroffes, blue petrels,
" and fheerwaters." *Cook.*—" During our run among the ice iflands,
                                                                    " the

" the pintadoes and the sheerwaters occurred in smaller numbers,
" but the penguins began to appear." *Cook.*—" As the weather
" was often calm, Mr. Banks went into the boat, to shoot birds, and
" he brought some albatrosses and sheerwaters; the latter were
" smaller than those which we saw in the straits of Le Maire, and had
" a deeper colour on the back." *Cook's First Voyage.*—" Sheerwaters
" are seen along the coast of Chili." *Carteret.*

[A] Specific character of the Black Skimmer, *Rynchops Nigra:*
" It is blackish, below white, its bill red at the base."

The N O D D Y.

Sterna Stolida. Linn. and Gmel.
Gavia Fusca. Briff.
Passer Stultus. Nieremb. Johnst. Will. and Charl.

MAN, who rules with haughty sway on land, is scarce known in another great division of nature's vast empire. On the stormy face of the seas, he finds enemies of superior force, obstacles that baffle his art, and dangers that exceed his courage. When he dares to pass those barriers of the world, all the elements combine to punish his audacity, and nature reclaims that dominion which he vainly aspires to usurp; there he is a fugitive, not a master. If he disturbs the inhabitants, if he ensnares or transfixes some unhappy victims, the bulk of them, safe in the bosom of the abyss, will in some future period see the winds and storms, and piercing colds, sweep from the face of the ocean its troublesome and destructive guests.

In fact the animals, which nature, though with feebler faculties, has fortified against the billows and the tempests, know not our dominion. Most of the sea-birds suffer us to approach them, and even to seize them, with a degree of unconcern that appears to border on stupidity, but which clearly

THE NODDY.

clearly evinces that man is to them a new and
ftrange being, and that, far removed from his
controul, they enjoy full and entire liberty. We
have already feen feveral inftances of that apparent weaknefs, or rather profound fecurity, which
characterizes the winged inhabitants of the ocean.
The Noddy, of which we now treat, has been
termed the *foolifh fparrow (paffer ftultus)*; a
very inaccurate denomination, fince the Noddy
is not a fparrow, but refembles a tern or little
mew, and, in fact, forms the gradation between
thefe two kinds of birds: for it has the feet of the
mew, and the bill of the tern. All its plumage
is dark brown, except a white fpace, like a cowl,
on the top of the head. Its bulk is nearly equal
to that of the common tern.

We have adopted the name *Noddy* (Noddi)
which occurs frequently in Englifh voyages *,
becaufe it expreffes the ftupidity, or filly confidence, with which the bird alights on the
mafts and yards of fhips † and even on the failors'
hands ‡.

* Particularly in thofe of Dampier and Cook.

† Thefe are ftupid birds, which, like the boobies, allow themfelves to be caught by the hand, on the yards and the rigging of the
veffel, on which they alight. *Catefby.*

‡ The *Thouaroux* (the name of the Noddies in Cayenne) come
to fifh on very ample fpace, in company with the frigats; I never
faw them alight on the water, like the gulls; but at night they come
roving about the veffels to find repofe, and the failors catch them by
lying on the top of the ftern, and ftretching out their hand, upon
which the birds make no fcruple to alight. *Memoirs communicated
by M. de la Borde, king's phyfician at Cayenne.*

The

The species seems not to extend much beyond the tropics *; but is very numerous in
its haunts. " At Cayenne," says M. de la
Borde, " there are an hundred Noddies for one
" booby, or man-of-war bird : they particularly
" cover the rock of the *Grand Connetable*,
" whence they come to fly about the vessels;
" and when a cannon is fired, they rise embo-
" died in a thick cloud." Catesby also saw
them in great numbers, flying together, and
dropping continually on the surface of the sea,
to catch the little fish, shoals of which are
impelled by violent winds. The birds seem to
perform their part with great alacrity and cheer-
fulness, if we judge from the variety of their
cries, and their great noise, which may be heard
some miles. " All this," adds Catesby, " has
" place only in the breeding season, when they
" lay their eggs on the naked rock †: after
                                    " which

Catesby.—Noddies and egg-birds (which seem to be a kind of
tern) in 27° 4′ south latitude, and 103° 56′ west longitude, about
the beginning of March. *Cook.*—On the 28th February, in 33° 7′
south latitude, and 102° 33′ west longitude (in sailing towards the
tropic) we began to see flying fish, egg-birds, and Noddies, which
are said not to go above sixty or eighty leagues from land; but of
this we have no certainty. No one yet knows to what distance any of
the oceanic birds go to sea; for my own part, I do not believe that
there is one in the whole tribe that can be relied on, in pointing out
the vicinity of land. *Idem.*—The Noddies are seen more than an
hundred leagues from land. *Catesby.* (The egg-bird of Cook is
the same with the Noddy of Dampier, and is the sooty tern, *Sterna
Fuliginosa*, already described).

‡† On those of Bahama. *Catesby.*—On the isle of Rocca. *Dampier.*—

" which each Noddy ranges at large, and roves
" folitary on the vaft ocean."

pier.—On the fouth fide of St. Helena, by feveral fmall iflets, which are properly but rocks, where we fee thoufands of black gulls, whofe eggs, which are very good eating, were laid on the bare rock. The multitude of thefe birds was fuch, that we took thoufands of them, and they fuffered themfelves to be knocked down with fticks; whence, no doubt, they have been called *foolifh gulls*. *Recueil des Voyages de la Compagnie des Indes Orientales.*

[A] Specific character of the Noddy, *Sterna Stolida:* " Its body " is black, its front whitifh, its eye-brows intenfe black."

The AVOSET.

L' AVOCETTE. *Buff.*

Recurviroſtra-Avoſetta. Linn. and Gmel.
Avocetta. Geſner, Aldrov. Johnſt. Will. Briſſ. &c.
Recurviroſtra. Rzacynſki, Barrere, &c.
Plotus Recurviroſter. Klein.
The Scooper. Charleton.
The Crooked-bill. Dale and Plott.
The Scooping Avoſet. Penn. and Lath. *

THE webbed birds have, for the moſt part, ſhort legs. Thoſe of the Avoſet are very long; and this diſproportion, which would almoſt alone diſtinguiſh it, is attended with a character ſtill more ſingular, that is, the inverſion of its bill, which is bent into an arc of a circle, whoſe centre lies above the head: the ſubſtance of the bill is ſoft and almoſt membranous at its tip †; it is thin, weak, ſlender, compreſſed horizontally, and incapable of defence and effort. It is one of

---

* The word *Avocetta* is of Italian origin; the bird has alſo in Italy the names *Beccotorto* and *Beccorella,* expreſſive of its crooked bill; and on Lake Maggiore it is called *Spinzago d' Aqua,* to diſtinguiſh it from the curlew, which is termed ſimply *Spinzago.* In Germany it is ſtyled *Frembder Waſſer Vogel* (foreign water bird), and *Schabel* or *Schnabel*; and in Auſtria *Krambſchabl*: in Sweden *Skiaër-ſlaëcka*: in Daniſh *Klyde, Lanfugl, Forkeert, Reguſpove*: in Turkey *Zeluk* or *Keluk.*

† Linnæus.

thoſe

THE SCOOPING AVOSET.

thofe errors or effays of nature, which if carried
a little farther would deftroy itfelf; for if the curvature of the bill were a degree increafed, the
bird could not procure any fort of food, and the
organ deftined for the fupport of life would infallibly occafion its deftruction. The bill of the
Avofet may therefore be regarded as the extreme
model which nature could trace, or at leaft preferve; and for that reafon it is the moft diftant
from the forms exhibited in other birds.

It is even difficult to conceive how this bird
feeds by help of an inftrument that can neither
peck nor feize its prey, but only rake in the
fofteft mud. It feems to employ itfelf in fearching the froth of the waves for fifh-fpawn, which
appears to be its chief fupport. It probably eats
worms alfo; for its bowels contain a glutinous
fubftance, fat to the touch, of a colour bordering
on orange yellow, in which are fome veftiges of
fifh-fpawn and aquatic infects. This gelatinous
mafs is always mixed in the ftomach with little
white cryftalline ftones*: fometimes in the
inteftines there occurs a gray or earthy green
matter, which feems to be the flimy fediment
which frefh waters, fwelled by rains, depofit on
their bed. The Avofet frequents the mouths of
rivers and ftreams†, in preference to other parts
of the fea-fhore.

* Willughby fays, that he could find nothing elfe.
† At leaft in Picardy, where thefe obfervations were made.
(In England alfo, at the mouth of the Severn.)

This

This bird is fomewhat larger than the lapwing: its legs are feven or eight inches high;
its neck is long, and its head round; its plumage is fnowy white on all the fore fide of the
body, and interfected with black on the back;
the tail is white, the bill black, and the feet
blue.

The Avofet runs by means of its long legs on
bottoms covered with five or fix inches of water:
but in deeper parts, it fwims, and in all its motions it appears lively, alert, and volatile. It
ftays but a fhort while in the fame place; and
in its paffages to the coaft of Picardy, in April
and November, it often difappears the morning
after its arrival: fo that fportfmen find great difficulty to kill or catch a few. They are ftill
more rare in the inland country: yet Salerne fays,
that they have been feen to advance pretty high
on the Loire. He affures us, that they are very
numerous on the coafts of Low Poitou, where
they breed *.

It appears from the route which the Avofets
hold in their paffage, that, on the approach of
winter, they journey towards the fouth, and return in the fpring to the north: for they occur
in Denmark †, in Sweden, on the fouthern

---

* The Avofet is very rare in the Orleanois ... On the contrary,
nothing is more common on the coafts of Lower Poitou; and in
the breeding feafon, the peafants take their eggs by thoufands to
eat: when driven off its neft, it counterfeits lamenefs as much or
more than any other bird. *Salerne.*

† Muller and Brunnich.

point

point of the ifle of Oëland*, on the eaftern
coafts of Great Britain†. Flocks of them arrive
alfo on the weftern fhore of that ifland, but remain no longer than a month or two, and retire
when the cold fets in ‡. Thefe birds only vifit
Pruffia ‖; they very feldom appear in Sweden;
and, according to Aldrovandus, they are not
more frequent in Italy, though well known there,
and juftly named §. Some fowlers have affured
us, that their cry may be expreffed by the fyl-
lables, *crex, crex*. But we cannot, on fuch flender authority, infer, that the Avofet is the fame
with the crex of Ariftotle:—" For the *crex*,"
fays the philofopher, " *wages war againft the
" oriole and the blackbird.* And the Avofet can
certainly have no quarrel with two birds which
inhabit the woods. Befides, the cry, *crex, crex,*
belongs equally to the jaducka fnipe and the
land rail.

In moft of the Avofets there is dirt on
the rump, and the feathers feem worn off by
rubbing. Probably thefe birds wipe their
bill on their feathers, or lodge it among them
when they fleep; fince the form feems as cum-
berfome to be difpofed during reft, as awkward
for action, unlefs, like the pigeon, it lays its
head on its breaft during repofe.

* Linnæus.
† Ray.
‡ Charleton.
‖ Rzaczynfki.
§ *Beccotorto*; i. e. twifted-bill.

The

The obferver * who has communicated thefe facts is perfuaded, that the Avofet is at firft gray, and he adopts this opinion becaufe many of thofe which arrive in November have the tips of their fcapular feathers gray, as well as thofe of the rump : but thefe feathers, and thofe which cover the wings, preferve longeft the livery of their birth : the dull colour of the great quills of the wings, and the pale tint of their legs, which in the adults are of a fine blue, leave no doubt but the Avofets whofe plumage is mixed with gray are young ones. There are few exterior differences in this fpecies between the male and female : the old males have much black, but the old females have nearly the fame ; only the latter feem to be fmaller, and the head of the former rounder, with the flefhy tubercle that rifes under the fkin, near the eye, more inflated. We ought not to admit varieties into the fpecies, though the Avofets of Sweden, according to Linnæus, have the rump black, which is white in the multitudes that inhabit a certain lake in Lower Auftria, as Kramer remarks.

Whether from timidity or addrefs, the Avofet fhuns fnares, and is very difficult to take †. The fpecies is no where common, and feems to contain few individuals.

* M. Baillon, of Montreuil-fur-mer.

† " I have practifed every poffible ftratagem to take thefe birds, " but could never fucceed." *Obfervations communicated by M. Baillon.*

[A] Specific character of the Scooping Avoset, *Recurviroftra-Avocetta:* " It is variegated with black and white." The Avofets are frequent on the eaftern fhores of Great Britain in winter; they alfo vifit the mouth of the Severn, and fometimes the pools of Shropfhire. They feed on the worms and infects which they fcoop out of the fand, which often fhews the marks of their bill. They lay two white eggs, as large as a pigeon's, of a greenifh hue, with large fpots of black. They are common in Tartary, about the Cafpian fea.

The R U N N E R.

Le Coureur. *Buff*.

Corrira Italica. Gmel.
Corrira. Briff.
Trochilus. Aldrov. Johnft. Will. Ray, and Charleton.
The Italian Courier. Lath.

ALL the birds which fwim, and whofe toes are connected by membranes, have the leg fhort, the thigh contracted, and often partly concealed under the belly. Their feet conftructed and difpofed like broad oars, with a fhort handle, and in an oblique pofition, feem exprefsly calculated for impelling the little animated fhip: the bird is at once the veffel, the rudder, and the pilot. But amidft this grand fleet of winged navigators, three fpecies form a feparate fquadron: their feet are indeed furnifhed with membranes like the other fwimming birds, but they are at the fame time raifed on tall legs, and, in this refpect, refemble the waders. Thus they form the intermediate gradation between two very different claffes.

Thefe three birds with tall legs and palmated feet are the avofet, the flamingo, and the Runner, fo called according to Aldrovandus, becaufe

caufe it runs fwiftly along the fhores. That naturalift, to whom alone we are indebted for the
account of this bird, tells us, that it is not rare
in Italy. But it is unknown in France, and in
all probability it occurs in no other country of
Europe, at leaft it is very uncommon. Charleton fays, that he faw one, without mentioning
whence it came. According to Aldrovandus,
the thighs of this bird are fhort in proportion
to the length of its legs: the bill is yellow
throughout, but black at the tip; it is fhort, and
does not open much: the mantle is iron gray,
and the belly white: two white feathers with
black points cover the tail. This is all that the
naturalift informs us; he adds nothing about
its meafures; but, if we judge from his figure,
they are nearly the fame with thofe of the
plover.

Both Ariftotle and Athenæus fpeak of a bird
that runs fwiftly, and which they term *trochilos*,
faying, that " it comes in calm weather, to feek
" its food on the water." But is this bird a
palmiped and fwimmer, as Aldrovandus afferts,
while he refers to it his courier; or is it not, as
Ælian hints, a wader of the kind of gallinules or
ringed plovers? It feems difficult to decide, from
the fcanty information tranfmitted from the ancients. All that we can gather is, that this
trochilos is an aquatic bird; and with fome probability Ælian refers to it the report of antiquity, that it entered boldly the jaws of the
crocodile

crocodile to eat the leeches, and warn it of the approach of the *ichneumon*. This fable has been applied the moſt abſurdly imaginable to the gold-creſted wren, from a confuſion of names, that little choriſter being often termed *trochilos*, becauſe of its whirling flight.

The RED FLAMINGO.

LE FLAMMANT, *ou* LE PHENICOPTERE.

Buff.

Phœnicopterus Ruber. Linn. and Gmel.
Phœnicopterus. Gefner, Aldrov. Johnft. Ray, Briff. &c. *

THE name *Phœnicopterus,* applied by the Greeks, and adopted by the Romans, expreffes the ftriking feature of this bird—the crimfon colour of its wings. But this is not the only remarkable character of the bird: the bill is flattened with a fudden bend above, thick and fquared below, like a broad fpoon; its legs are exceffively tall; its neck is long and flender; its body is more elevated, though fmaller, than that of the ftork, and prefents a fingular and confpicuous figure among the great waders.

* In Greek Φοινικοπτερος, from φοινιξ, the Phœnician dye, and πτερον, a wing. And hence this name has, in the modern languages, been tranflated by words denoting flame-colour. In Portuguefe *Flamingo :* in Spanifh *Flamenco :* in French *Flambant* or *Flammant ;* which, as Buffon fays in the text (we have omitted the paffage) was afterwards written *Flamand (Flemifh) ;* and, by this ridiculous miftake, the bird was imagined to be a native of Flanders. In France it was anciently called *Becharu,* becaufe its bill refembles a plough-fhare *(foc de charrue).* In Cayenne it has the name *Tokoko.*

Thofe

Thofe large femi-palmated birds, which haunt the fides of waters, but neither dive nor fwim, are judicioufly reckoned by Willughby diftinct and independent fpecies: for the Flamingo in particular feems to form the gradation between them and the clafs of the great fwimmers, which it refembles by its half-webs, and becaufe the membrane ftretched between the toes recedes in the middle by two fcallops. All the toes are very fhort, and the outer one extremely little: the body alfo is fmall in proportion to the length of its wings and neck. Scaliger compares it to that of the heron, and Gefner to that of the ftork; remarking, as well as Willughby, the exceffive length of the flender neck. " When the Flamingo has attained its full " growth," fays Catefby, " it is not heavier " than a wild duck, and is yet five feet high." Thefe great differences in fize, noticed by authors, have a reference to the age as well as to the varieties which they have alfo remarked in the plumage. This is generally foft and filky, and wafhed with red tints of greater or lefs vivacity and extent: the great quills of the wing are conftantly black: the coverts, both the greater and leffer, the exterior and interior, are imbued with fine flame-colour; which fpreads and dilutes by degrees over the back and the rump, the breaft and the neck; on the upper part of which, and on the head, the plumage is a

fhaven

shaven and velvet down. The top of the head
is naked; the neck is very slender, and the bill
is broad; so that the bird has an uncommon appearance. Its skull seems to be raised and its
throat dilated before, to receive the lower mandible, which is very broad at its origin. The
two mandibles form a round and straight canal
as far as their middle; after which the upper
one bends suddenly, and its convexity changes
into a broad surface: the lower mandible reflects
proportionally, but always preserves the shape
of a broad gutter; and the upper one, by a
small curvature at its point, applies to the extremity of the lower mandible. The sides of
both are beset internally with a small black indenting, whose points are turned backwards.
Dr. Grew, who has described this bill with
great accuracy, remarks also a filament within,
under the upper mandible, and which divides it
in the middle. It is black from its tip to where
it bends, and from thence to the root it is white
in the dead bird, but, in the living subject, it
seems liable to vary; since Gesner asserts, that
it is of a bright red, Aldrovandus that it is brown,
Willughby that it is blueish, and Seba that it is
yellow. " To a small round head," says Dutertre, " is joined a large bill four inches long,
" half red, half black, and bent into the form of
" a spoon." The Academicians, who have described this bird under the name of *Becharu,*

say,

fay, that its bill is of a pale red, and contains a
thick tongue edged with flefhy *papillæ*, turned
backwards, which fills the cavity or the large
fpoon of the lower mandible. Wormius alfo
defcribes this extraordinary bill; and Aldrovandus remarks how much nature has fported in
its conformation: Ray fpeaks of its ftrange
figure. But none of them have examined it with
fuch attention as to decide a point which we
fhould be glad to afcertain, viz. whether, as
many naturalifts alledge, the upper mandible is
moveable, while the lower is fixed.

Of two figures of this bird, publifhed by Aldrovandus, and fent to him from Sardinia, the
one exprefles not the characters of the bill, which
are accurately portrayed in the other. And we
muft remark by the way, that in our plate
the fwelling and flattening of the bill are too
faint, and that it is reprefented too much
pointed.

Pliny feems to clafs this bird with the ftorks,
and Seba has injudicioufly fuppofed that the
phœnicopterus was ranked by the ancients with
the ibis. But it belongs to neither of thefe
kinds: it forms a feparate divifion. And befides, when the ancients placed together analogous fpecies, they did not follow the narrow
views, or adhere to the fcholaftic methods, of
our nomenclators; they obferved in nature certain refemblances of habits and faculties, which
they conjoined in the fame group.

8   We

We may reafonably wonder that the name
phœnicopterus occurs not in Ariftotle, though
mentioned by Ariftophanes, who ranges it among
the marfh birds *. But it was rare and perhaps
foreign in Greece. Heliodorus † exprefsly
fays, that the *phœnicopterus* inhabited the Nile:
the old fcholiaft on Juvenal ‡ afferts, that it was
frequent in Africa. Yet thefe birds feem not
to remain conftantly in the hotteft climates; for
fome are found in Italy, and a much greater number in Spain §. It is only a few years fince feveral of them arrived on the coafts of Lan-
guedoc and Provence, particularly near Mont-
pellier and Martigues ‖, and in the fens near
Arles ¶. I am therefore aftonifhed that fo well-
informed an obferver as Belon fhould affert, that
none are ever feen in France, but fuch as had
been carried thither. Did this bird extend its
migrations firft to Italy, where it was anciently
foreign, and thence to the French coafts?

It inhabits, we fee, the countries of the fouth,
and is found from the coafts of the Mediterranean to the extremity of Africa **. Great

* Λιμναιος.
† Æthiopic. *lib.* vi.
‡ Sat. xi. 139.
§ Belon.
‖ Lifter, *Annot. in Apicium, lib.* v. 7.—Ray, *Synopf. p.* 117.
¶ Peirefc. vita, *lib.* ii.
** Thefe birds are very common at the Cape; they pafs the
day on the fides of the lakes and rivers, and at night they retire to
the mountains. *Kolben.*

numbers

numbers occur in the Cape de Verd iflands, according to Mandeflo, who over-rates the bulk of their body when he compares it to that of a fwan. Dampier met with fome nefts of thefe birds in the ifle of Sal. They are abundant in the weftern provinces of Africa, at Angola, Congo, and Biffao, where, from a fuperftitious refpect, the negroes will not fuffer one of them to be hurt *; and they live undifturbed in the midft even of the dwellings. They occur likewife in the bay of Saldana †, and in all the countries adjacent to the Cape of Good Hope, where they fpend the day on the coaft, and retire in the evening to the rank herbs which grow on fome parts of the contiguous lands ‡.

* The Flamingos are numerous in this canton, and fo refpected by the Mandingos of a village diftant half a league from Geves, that they are found in thoufands; thefe birds are of the bulk of a turkey-cock . . . the inhabitants of the fame village carry fo far the refpect for them, that they will not permit them to receive the leaft injury. They leave them tranquil on the trees amidft their dwellings, without being incommoded by their cries, which however are heard a quarter of a league. The French having killed fome of them in this afylum, were obliged to conceal them under the grafs, left the negroes fhould be prompted to revenge the death of a bird fo revered. *Relation de Brue, Hift. Gen. des Voy. tom.* ii. *p.* 590.

† In the multitude of birds feen in the bay of Saldana, the pelicans, the Flamingos, the ravens, which all have a white collar round the neck, numbers of fmall birds of different kinds, not to mention fea-fowl, which are of endlefs variety, fill the air, the trees, and the land, to fuch degree, that a perfon cannot ftir without putting up many. *Relation de Dounton, Hift. Gen. des Voy. tom.* ii. *p.* 46.

‡ Hift. Gen. des Voy. *tom.* v. *p.* 201.

The

The Flamingo is undoubtedly a migratory bird, but vifits only the warm and temperate regions, and never penetrates to the northern tracts. In certain feafons, they appear in feveral places, nor can we be certain whence they come, but they never feem to travel towards the north; and if fome folitary ftragglers are found at times in the interior parts of France, they have been driven thither in a ftorm. Salerne relates, as an extraordinary occurrence, that one was killed on the Loire. The hot countries are the fcene of their migrations: and they have traverfed the Atlantic; for they are of the fmall number of birds that inhabit the tropical regions of both continents *.

They are feen in Valparaifo, at Conception, and at Cuba †, where the Spaniards call them *flamencos* ‡. They occur on the coaft of Vene-

* In the ifland of Mauritius, or of France, there are many of the birds called *giants*, becaufe they carry their head fix feet high; they are exceeding tall, and their neck is very long; their body is not larger than that of a goofe: they feed in marfhy places; and the dogs often take them by furprize, as they require a confiderable time to rife from the ground. We once faw one at Rodrigue, and it was fo fat that we caught it with the hand: it is the only one which we ever remarked, which makes me think that it had been driven thither by fome violent wind, which it could not refift. This game is pretty good. *Leguat.*

† In the fmall iflands under Cuba, which Columbus called the *Queen's Garden*, there are red birds fhaped like cranes, which are peculiar to thefe iflands, where they live on falt-water, or rather on what they find proper for their fupport in it. *Herrera.*

‡ De Laet.

zuela,

zuela, near the White ifland, that of *Aves*, and that of *Roche*, which is a group of rocks *. They are well known at Cayenne, where the natives of the country name them *tococo* ; they fly in flocks on the fea-beach †. They inhabit alfo the Bahama iflands ‡. Sir Hans Sloane ranks them among the birds of Jamaica §. Dampier found them at Rio de la Hacha ‖. They are extremely numerous at St. Domingo ¶, in the Antilles and the Caribbee iflands **, where

* De Laet.

† Barrere.—The woods at Cayenne are inhabited by Flamingos, colibris, ocos, and toucans. *Voyage de Froger.*

‡ Klein.

§ " Thefe are common in the marfhy and fenny places and " likewife fhallow bays of Jamaica."

‖ I have feen Flamingos at Rio de la Hacha, and at an ifland fituated near the continent of America, oppofite to Curaçoa, and which the pirates call the *Flamingo Ifland*, becaufe of the prodigious number of thefe birds which breed in it. *Dampier.*

¶ In St. Domingo, the Flamingos appear in great numbers on the fkirts of the marfhes ; and as their feet are exceedingly tall, they may be taken at a diftance for an army in martial array. *Hift. Gen. des Voy. tom.* xii. *p.* 228.—The places which the Flamingos frequent the moft in St. Domingo, are the marfhes of Gonave and Cow Ifland, *(Ifle a Vache)* fmall iflands fituated, the one weft of Port-au-Prince, the other fouth of the city of Cayes. They are fond of thefe iflands, becaufe they are not inhabited, and becaufe they find in them many lagoons and falt-marfhes ; they alfo much frequent the famous pool of Riquille, which belongs to the Spaniards. They are feen eaft from the plain *Cul-de-fac,* in a great pool which contains many iflets ; but the number of thefe birds is obferved to diminifh in proportion as the marfhes are drained, and the tall timber cut down which fkirts them. *Extract from the Memoirs communicated by the Chevalier Lefebvre Defhayes.*

** Hernandez, Rochefort.

they

they live in the little falt pools and the lagoons.
That figured by Seba was fent him from Cura-
çoa. They occur alfo in Peru *, and as far as
Chili †. In fhort, there are few parts of South
America where navigators have not met with
them.

Thefe American Flamingos are entirely the
fame with thofe of Europe and Africa. The
fpecies appears fingle and disjoined, fince it
admits of no variety.

Thefe birds breed on the coafts of Cuba and
of the Bahama iflands ‡ on the deluged fhores,
and the low iflets, fuch as that of *Aves* §, where
Labat found a number with their nefts. Thefe
are little heaps of clayey and miry foil gathered
from the marfhes, and raifed about twenty inches
into a pyramid in the middle of the water, which
conftantly wafhes the bafe; the top is truncated, hollow, and fmooth, and, without any bed
of feathers or herbs, receives the eggs, which
the bird covers, fays Catefby, by fitting acrofs
the hillock ||, its legs hanging down, like a man

on

---

* De Laet.
† Frezier.
‡ Catefby.
§ Fifty leagues to the windward of Dominica.
|| I was fhown a great number of thefe nefts; they refemble
truncated cones, compofed of fat earth, about eighteen or twenty
inches high, and as much in diameter at the bafe; they are always
in water, that is, in meres or marfhes: thefe cones are folid to the
height of the water, and then hollow like a pot bored at top; in
this they lay two eggs, which they hatch by refting on them, and

covering

on a ftool: fo that only the rump and lower belly are of fervice in the incubation. This fingular pofition it is obliged to adopt on account of the length of its legs, which could never be bent under it if it were fquat. Dampier gives the fame defcription of the mode of hatching in the ifland of Sal *.

The nefts are always placed in the faltmarfhes; they contain only two or at moft three eggs, which are white, as thick as thofe of a goofe, and fomewhat longer. The young do not begin to fly till they have gained almoft their full growth; but they run remarkably fwiftly a few days after they are hatched.

The plumage is at firft of a light gray, and that colour becomes deeper, in proportion as

covering the hole with their tail. I broke fome, but found neither feathers, nor herbs, nor any thing that might receive the eggs: the bottom is fomewhat concave, and the fides are very even. *Labat.*

* They make their neft in the marfhes, where they can find plenty of flime, which they heap with their claws, and form hillocks refembling little iflets, and which appear a foot and a half above the water; they make the bafe broad, and taper the ftructure gradually to the top, where they leave a fmall hollow to receive their eggs. When they lay or hatch, they ftand erect, not on the top but very near it, their feet on the ground and in the water, leaning themfelves againft the hillock, and covering the neft with their tail: their legs are very long, and as they make their neft on the ground, they could not, without injuring their eggs or their young, have their legs in the neft, nor fit, nor fupport their whole body but for this wonderful inftinct which nature has given them. They never lay more than two eggs, and feldom fewer. Their young ones cannot fly till they are almoft full grown; but will run prodigioufly faft. *Dampier.*

their

their feathers grow; but it requires ten or twelve months before their body attains its full fize, and then they affume their fine colour, whofe tints are faint when they are young, and grow deeper and brighter as they advance in age *. According to Catefby, two years pafs before they acquire the whole of their beautiful red colour. Father Dutertre makes the fame remark †. But whatever be the progrefs of this tint in the plumage, the wing firft acquires the colour, and is always brighter than the other parts. The red afterwards fpreads from the wing to the rump, then to the back and the breaft, and as far as the neck: only in fome individuals there are flight varieties of fhades, which feem to follow the differences of climate: for example, the Flamingo of Senegal feemed to have a deep red, and that of Cayenne inclined to orange; but that variation was not enough to conftitute two fpecies, as Barrere has done.

Their food is in every country nearly the fame. They eat fhell-fifh, fifh-fpawn, and aquatic infects: they feek them in the mud, into which they thruft their bill and part of their head; at

---

* They differ in colour, their plumage being white when they are young; then, in proportion as they grow, they become rofe coloured; and laftly, when aged, they are entirely carnation. *De Laet, and Labat.*

† " The young are much whiter than the old ones; they " incline to red as they grow old: I have feen fome alfo which had " their wings mixed with red, black, and white feathers; I believe " that thefe are males.

the

the fame time they continually pufh their feet
downwards, to carry the prey with the flime to
their bill, which is fitted by its indenting to retain any fubftance. It is a fmall feed, fays
Catefby, like millet, that they bring up by thus
puddling in the mire. But it probably is nothing
elfe than the egg of fome infect; for the flies
and gnats are furely as abundant in the overflowed plains of America as in the low grounds
of the north, where Maupertuis faw whole
lakes covered with fuch eggs, refembling the
grains of millet *. In the iflands of the new
world, thefe birds may find abundance of this
fort of food; but on the coafts of Europe they
fubfift on fifh, the indentings of their bill ferv-
ing like teeth to hold the flippery prey.

They appear to prefer the fea-fhore: if they
are feen on rivers, fuch as the Rhone †, it is
never far from their mouth. They haunt more
conftantly the inlets, falt-marfhes, and low
coafts; and it has been obferved, that in rear-
ing them they require falt-water to drink ‡.

Thefe birds always go in flocks; and to fifh,
they naturally form themfelves into a line, which
at a diftance has a fingular appearance, like a
file of foldiers §. This propenfity to difpofe
themfelves in ranks ftill adheres to them when,

* Œuvres de Maupertuis, *tom.* iii. *p.* 116.
† Peirefc. vita, *lib.* ii.
‡ De Laet, Labat, and Charlevoix.
§ Hift. Gen. des Voy. *tom.* xii. *p.* 229.

placed

placed one againſt another, they repoſe on the
beach *. They ſtation ſentinels and keep a ſort
of guard, as uſual with all gregarious birds.
When they are engaged in fiſhing, their head
plunged in the water, one of them remains
ſentry, keeping his head erect † : on the leaſt
menace of danger, he gives a loud cry, audi-
ble at a great diſtance, and much like the ſound
of a trumpet ‡ ; inſtantly the whole flock riſes,
and preſerves in its flight an order ſimilar to that
of cranes. Yet if theſe birds be ſuddenly ſurprized, they remain ſtupid and motionleſs thro'
fear, and afford the fowler time to knock them

* They uſually reſt upon their legs, one againſt the other, in a
ſingle line ; in this ſituation any perſon at the diſtance of half a
mile would take them for a brick wall, becauſe they have exactly
the ſame colour. *Roberts, Hiſt. Gen. des Voy. tom.* xi. *p.* 364.

† " They are conſtantly on their guard againſt any ſurprizal by
" their enemies, and it is alledged that ſome ſtand as ſentinels while
" the reſt are occupied in ſearching for food ; beſides, they are ſaid
" to ſmell powder at a diſtance, and are therefore approached with
" difficulty. Our old buccaneers employed a ſtratagem for killing
" them ſimilar to what the people of Florida are ſaid to uſe, in
" order to approach the deer ; they covered themſelves with ox-
" ſkins, and advanced againſt the wind upon the Flamingos, which,
" being accuſtomed to ſee oxen feed in the ſavannas, are not intimi-
" dated, and thus the hunters can eaſily fire at them." *Charlevoix.*

‡ " Theſe birds have ſuch a ſtrong voice, that any perſon hear-
" ing them would ſuppoſe they were trumpets ſounding ; and while
" they have their head concealed, dabbling in the water, like the
" ſwans, to find there ſubſiſtence, there is always one that continues
" erect as ſentinel, its neck extended, its eye watchful, its head
" roving : as ſoon as it perceives a perſon, it ſounds the trumpet,
" gives alarm to its diſtrict, riſes the firſt on wing, and all the
" reſt follow it." *Idem.*

down

down one after another. Of this we are informed
by Dutertre, and it may alfo reconcile the oppofite accounts of navigators; fome reprefenting
the Flamingos as timorous birds *, which can
hardly be approached †, while others affert, that
they are heavy and ftupid ‡, and fuffer themfelves
to be killed one after another ‖.

Their flefh is highly efteemed. Catefby compares its delicacy to that of the partridge: Dampier fays, that it has a very good flavour, though
lean: Dutertre found it to be excellent, notwithftanding a flight marfhy tafte. Moft travellers
give the fame account §. M. de Peirefc is almoft
the only one who afferts that it is bad: but, befides the difference produced by climate, thefe

* " Their hearing and fmell are fo acute, that they can wind at
" a great diftance the fowlers and the fire-arms; and alfo, to avoid
" all furprize, they prefer alighting on open places in the midft
" of marfhes, whence they can defcry their enemies from afar, and
" there is always one of the band that keeps watch." Rochefort,
Hift. des Antilles.

† Thefe birds are difficult to approach: Dampier and two other
fowlers having placed themfelves in the evening near their retreat,
furprized them fo fuccefsfully as to kill fourteen at three fhots.
Roberts, in the *Hift. Gen. des Voy. tom.* ii. *p.* 364.

‡ " *Stolida Avis,*" fays Klein.

‖ A man concealing himfelf from their fight, may kill a great
number of them; for the report of a difcharge does not make them
ftir, nor are they alarmed at feeing their companions killed in the
midft of them; but they remain with their eyes fixed, and, as it
were, ftruck with aftonifhment, till they are all deftroyed, or at leaft
moft of them. *Catefby.*

§ Thefe birds are numerous near the Cape; their flefh is wholefome and favoury; their tongue is faid to have the tafte of marrow.
Hift. Gen. des Voy, tom. v. *p.* 201.—They are fat, and their flefh is
delicate. *Rochefort.*

birds

birds muſt be exhauſted and lean with fatigue,
when they arrive on our coaſts. The ancients
ſpeak of them as being exquiſite game *. Philo-
ſtratus reckons them among the delicacies of entertainments †. Juvenal, upbraiding the Romans
with their waſteful luxury, ſays, that they cover
their tables with the rare birds of Scythia, and
with the *phœnicopterus*. Apicius deſcribes the
ſcientific mode of ſeaſoning them ‡; and it was
this man, *who*, ſays Pliny, *was the deepeſt abyſs of*
waſtefulneſs ||, that diſcovered in the tongue of
the Flamingo that exquiſite reliſh, which recom-
mended it ſo highly to epicures §. Some of our
navigators,

---

* When Caligula had reached ſuch a pitch of folly as to fancy
himſelf a divinity, he choſe the *phœnicopterus* and the peacock as the
moſt exquiſite victims to be offered up to his godſhip; and the day
before he was maſſacred, ſays Suetonius, he was beſprinkled at a ſa-
crifice with the blood of a *phœnicopterus*. *In Vit. Calig. c.* 57.

† Vita Apollon. *lib.* viii.

‡ " Cleanſe, waſh, and truſs the *phœnicopterus*; put it into a
" kettle; add water, ſalt, and a little vinegar. At half boiling, tie
" in it a bunch of leeks and coriander, that it may ſtew : near boil-
" ing, drop into it ſpiced wine, and colour the mixture. Put into
" a mortar pepper, cummin, coriander, the root of laſer, mint,
" rue; pound theſe, pour on vinegar, add walnut-date. Pour
" on it its own gravy, and turn the whole back into the ſame ket-
" tle : cloſe it with ſtarch; pour on the gravy, and carry it in."
———*Otherwiſe :* " Roaſt the bird; grind pepper, loveage, par ſley-
" ſeeds, ſeſame, ſpiced wine, wild parſley, mint, dry onions,
" walnut-dates ; and temper the whole with honey, wine, pickle,
" vinegar, oil, and ſpiced wine." *De Obſon. & Condim.* lib. vi. 7.

|| *Phœnicopteri linguam præcipui eſſe ſaporis Apicius docuit, nepotum*
omnium altiſſimus gurges.

§ Lampridius reckons among the extravagancies of Heliogaba-
lus, his ordering for his table diſhes filled with the tongues of the
phœnicopterus.

navigators, whether from the prejudice derived from antiquity, or from their own experience, commend the delicacy of that morfel *.

The fkin of thefe birds, which is well clothed with down, ferves for the fame purpofes as that of the fwan †. They may be eafily tamed, either by taking them young from the neft ‡, or by enfnaring the adults in gins, or any other way ‖ ;

for

phœnicopterus. Suetonius fays, that Vitellius bringing together the delicacies of all the parts of the world, caufed to be ferved up at his entertainments, at once, the livers of fcari, the roes of muræna, the brains of pheafants and peacocks, and the tongues of *phœnicopters ;* and Martial, upbraiding the Romans for their deftructive tafte, makes this bird complain in the following lines :

> *Dat mihi penna rubens nomen ; fed lingua gulofis*
> *Noftra fapit : quid, fi garrula lingua foret ?*

* But above all, their tongue paffes for the moft exquifite morfel that can be eaten. *Dutertre.*—Their tongue is very large, and near the root there is a lump of fat, which makes an excellent morfel. A plate of Flamingos tongues, according to Dampier, would be a difh fit for the king's table. *Roberts.*

† They are flayed, and their fkins are made into excellent fur, which would be very ufeful for perfons troubled with a cold debilitated ftomach. *Dutertre.*

‡ I wifhed much to have young ones to tame ; for this fucceeds, and I have feen fome very familiar at the houfe of the governor of Martinico ... In lefs than four or five days the young ones which we took came to eat out of our hands ; yet I kept them always faftened, without trufting much to them ; for one which was loofened fled as fwift as a hare, and my dog could with difficulty overtake it. *Labat.*

‖ " A wild Flamingo having alighted in a mere near our dwel-
" ling, a tame Flamingo was driven thither, and the negro boy who
" had the charge of it, carried the trough in which it fed to the edge
" of the mere, at fome diftance, and concealed himfelf hard by :
" the tame Flamingo foon approached, and the wild one followed,
" and defiring to partake in the repaft, it began to fight and chafe
" its

for though very wild in the ftate of liberty, the
Flamingo, when once caught, is fubmiffive, and
even affectionate. In fact, it has rather a timorous than a lofty fpirit; and the fame fear which
prompts it to fly, fubdues it after it is taken.
The Indians have completely tamed them. M.
De Peirefc faw them very familiar, fince he gives
feveral particulars of their domeftic life. They
eat more in the night, he fays, than in the day,
and foak their bread in water. They are fenfible to cold, and creep fo clofe to the fire as to
burn their feet; and when one leg is difabled,

" its rival; fo that the little negro, who lay on the ground as if he
" had been dead, fnatched the opportunity to catch the bird by
" feizing its legs.   One of thefe Flamingos, caught nearly in the
" fame manner, lived fifteen years in our court-yard; it continued
" on good terms with the poultry, and even careffed its fellow-
" lodgers, the turkies and ducks, by fcratching their back with its
" bill.   It fed on the fame grain as the other poultry, provided that
" it was wetted with a little water; it could eat only by turning the
" bill to lay hold of its food fidewife: it dabbled like the ducks,
" and knew thofe perfons fo well who ufually took care of it, that
" when hungry it went to them and pulled their clothes with its
" bill: it often kept itfelf mid-legs in water, feldom changing its
" place, and plunging from time to time its head to the bottom, to
" catch fmall fifhes, which it prefers to grain.   Sometimes it ran
" on the water, ftriking it alternately with its claws, and fupport-
" ing itfelf by the motion of its wings half extended.   It was not
" fond of fwimming, but only of puddling with its feet in fhallow
" water. When it fell, it rofe with great difficulty; and accordingly
" it never refted on its belly to fleep: it only drew one of its legs
" under it, leaned upon the other, paffed its neck upon its back,
" and concealed its head between the end of its wing and its body,
" always on the fide oppofite to the leg which was bent." *Letter from*
M. Pommies, *commander of militia in the diftrict of Nipes, at St. Do-*
mingo, communicated by the Chevalier Defhayes.

they

they walk on the other, and affift their motion by ufing their bill like a crutch on the ground. They fleep little, and reft only on one leg, the other being drawn under the belly. Yet they are delicate, and difficult to rear in our climates: it appears even, that, though pliant to the habits of captivity, that ftate is very unfuitable to their nature, fince they cannot fupport it long, but drag out a languifhing exiftence; for they never propagate when reduced to domeftication.

END OF THE EIGHTH VOLUME.

Printed in the United States
By Bookmasters